Springer Theses

Recognizing Outstanding Ph.D. Research

For further volumes:
http://www.springer.com/series/8790

Aims and Scope

The series "Springer Theses" brings together a selection of the very best Ph.D. theses from around the world and across the physical sciences. Nominated and endorsed by two recognized specialists, each published volume has been selected for its scientific excellence and the high impact of its contents for the pertinent field of research. For greater accessibility to non-specialists, the published versions include an extended introduction, as well as a foreword by the student's supervisor explaining the special relevance of the work for the field. As a whole, the series will provide a valuable resource both for newcomers to the research fields described, and for other scientists seeking detailed background information on special questions. Finally, it provides an accredited documentation of the valuable contributions made by today's younger generation of scientists.

Theses are accepted into the series by invited nomination only and must fulfill all of the following criteria

- They must be written in good English.
- The topic should fall within the confines of Chemistry, Physics, Earth Sciences, Engineering and related interdisciplinary fields such as Materials, Nanoscience, Chemical Engineering, Complex Systems and Biophysics.
- The work reported in the thesis must represent a significant scientific advance.
- If the thesis includes previously published material, permission to reproduce this must be gained from the respective copyright holder.
- They must have been examined and passed during the 12 months prior to nomination.
- Each thesis should include a foreword by the supervisor outlining the significance of its content.
- The theses should have a clearly defined structure including an introduction accessible to scientists not expert in that particular field.

Paola Di Pietro

Optical Properties of Bismuth-Based Topological Insulators

Doctoral Thesis accepted by
the Technical University “Sapienza” of Rome, Italy

Springer

Author
Dr. Paola Di Pietro
INSTM UdR Trieste-ST Area Science Park
Trieste
Italy

Supervisors'
Prof. Dr. Paolo Calvani
Prof. Dr. Stefano Lupi
Dipartimento di Fisica and CNR-SPIN
Università di Roma "La Sapienza"
Piazzale Aldo Moro 2 00185
Rome
Italy

ISSN 2190-5053 ISSN 2190-5061 (electronic)
ISBN 978-3-319-01990-1 ISBN 978-3-319-01991-8 (eBook)
DOI 10.1007/978-3-319-01991-8
Springer Cham Heidelberg New York Dordrecht London

Library of Congress Control Number: 2013949665

Printed on acid-free paper

Springer is part of Springer Science+Business Media (www.springer.com)

Supervisors' Foreword

A few years ago a new field emerged in condensed matter physics and immediately attracted a multitude of theorists and experimentalists, that of Topological Insulators (TIs). At their interface with another insulator (like a vacuum), the special topology of the electronic bands under a strong spin–orbit interaction builds up a metallic layer of nanometric depth. The phenomenon is similar to the spin quantum Hall effect, but it occurs in three dimensions, with no external magnetic field, and even at room temperature. Therefore, at variance with the Dirac fermions of graphene, those of TIs appears spontaneously in crystals or films without the need to physically extract a monoatomic layer from the bulk material and to dope the layer. Moreover, in TIs the mechanism of spin-lockage makes their carriers extremely robust against backscattering processes, thus providing a very high sheet conductance.

In crystalline TIs, however, one has to face a major problem, namely the impurity concentration: the number of the electrons they inject into the crystal should be vanishingly small, if one wants to study the properties of Dirac fermions by using a bulk technique like infrared spectroscopy. The first set of results in the thesis of Dr. Di Pietro is devoted to optically identify, in a variety of different bismuth chalcogenides grown by the group of Robert Cava at the Princeton University, those which were the best compensated ones. Unfortunately, even the best compound still presented—also at low temperature—a residual conductivity comparable with that of the surface crystals.

Therefore, in the second part of the thesis this study has been extended to several thin films of Bi_2Se_3, grown at the Rutgers University by the group of S. Oh. The ratio surface electrons/bulk electrons could then be largely enhanced and this time the results were satisfactory. Dr. Di Pietro could then finally connect the fascinating field of TIs with Plasmonics, one of the most studied subjects in contemporary nanotechnology. Plasmons are the collective excitations of an electron gas, and they create mixed states, called surface plasmon-polaritons, when they interact with the electromagnetic radiation at the surface of metals. In order to detect these elusive quasiparticles, one has to implement onto the metal periodic structures having individual dimensions much smaller than the radiation wavelength. In the Terahertz domain, this can be achieved by patterning the conducting surface in form of ribbons of micrometric size.

This has been done by Dr. Di Pietro, in the second part of this experimental work, in order to first detect the surface plasmon-polaritons of topological insulators. The transmittance spectra of four thin films of different thickness and ribbon widths have clearly shown their absorption features and the analysis of their dispersion curve has confirmed not only the two-dimensional nature of the excitations but also, through further calculations, that they are associated with Dirac fermions.

This excellent result is not the only merit of the thesis written by Dr. Di Pietro. This work is also a very good example of how to approach a complex experimental problem, from a systematic characterization of the novel material and the choice of the samples, in view of the particular technique to be employed, to the identification of the experimental procedure and of features which can lead to an unambiguous demonstration of the phenomenon of interest. Therefore, we believe that the present thesis will be interesting for the reader and will stimulate further studies on the Topological Insulators and on their collective excitations.

August 2013
Rome, Italy

Prof. Dr. Paolo Calvani
Prof. Dr. Stefano Lupi

Contents

Abbreviations

TI	Topological insulator
IR	Infrared
FIR	Far Infrared
1(2, 3)D	One-(two, three) dimensional
2DEG	Two-dimensional electron gas
SS	Surface state
DP	Dirac point
ARPES	Angle resolved photoemission spectroscopy
FTIR	Fourier transform Infrared spectroscopy
QHE	Quantum Hall effect
IQHE	Integer quantum Hall effect
QSHE	Quantum spin Hall effect
TR	Time reversal symmetry
MOSFET	Metal-oxide-semiconductor field-effect-transistor
SdH	Shubnikov-deHaas
DOS	Density of states
TKNN	Thouless, Kohomoto, Nightingale, and den Nijis
SOC	Spin–orbit coupling
XRD	X-ray diffraction
QL	Quintuple layer
STM	Scanning tunneling microscope
XPS	X-ray photoemission spectroscopy
LDA	Local density approximation
FS	Fermi surface
BZ	Brillouin zone
CBM	Conduction band minimum
BCB	Bulk conduction band
BVB	Bulk valence band
TDTS	Time domain Terahertz spectroscopy
ZPD	Zero path difference
SR	Synchrotron radiation
CSR	Coherent synchrotron radiation
KK	Kramers–Kronig
SPP	Surface plasmon polariton

TM	Transverse magnetic
TE	Transverse electric
PEC	Perfect electric conductor
MBE	Molecular beam epitaxy
EBL	Electron beam litography
RIE	Reactive ion etching
RHEED	Reflection high-energy electron diffraction
SEM	Electron scan microscope
TO	Optic transverse normal mode
DS	Discrete state
CS	Continuum of states
PR	Plasmon resonance
IMT	Insulator-to-metal transition
D-L	Drude–Lorentz
SW	Spectral weight

Chapter 1
Introduction to the Topological Insulators and State of the Art

Abstract In this chapter a brief introduction to the world of topological condensed matter is presented. Firstly, the quantum Hall state in a two dimensional electron gas (2DEG) is introduced. After, the prediction and the subsequent experimental discovery of a topological insulating state of spin-orbit origin is presented. In particular, the so-called strong three dimensional topological insulators (3D TIs) are the object of this thesis. The second part of this chapter focuses on specific 3D TI materials. After their chemical description, their transport and surface properties are reported together with previous optical measurement results.

1.1 Introduction

In the 80s one of the most important experimental discoveries in condensed matter was that electrons confined in two dimensions and subject to a strong magnetic field display an original "topological" order, due to the so-called quantum Hall effect. In the last seven years it has been found that such a topological order also occurs in certain three-dimensional (3D) insulating materials. In these materials, the role of the external magnetic field is played by a strong spin-orbit coupling, an intrinsic property of all solids.

These materials have been called Topological Insulators (TIs) because they are insulators in the bulk, but have exotic metallic states at their surfaces, associated with a 2D electron gas (2DEG). This means that the topology, associated with the electronic wavefunctions of the system, changes when passing from the bulk to the surface.

The surface state (SS), characterized by those topological effects, makes the electron motion insensitive to scattering by non magnetic impurities. Such a dissipationless character may provide novel applications in technology, for instance in spintronics or quantum computing.

P. Di Pietro, *Optical Properties of Bismuth-Based Topological Insulators*,
Springer Theses, DOI: 10.1007/978-3-319-01991-8_1,

The early theoretical prediction and the first experimental observation of TIs have demonstrated that the 3D TIs sustain the spin quantum Hall effect and their edge provides a state protected by time reversal (TR) symmetry and backscattering. On the surface of a 3D TI the electron motion is allowed in any direction along the surface, but those directions uniquely determine the electron spin polarization and *viceversa*. The 2D energy-momentum relation provides a Dirac cone structure, whose particles are mass-less.

By means of Angle Resolved Photoemission (ARPES) some bismuth compounds have been demonstrated to be 3D TIs. In particular Bi_2Se_3 and Bi_2Te_3 exhibit a single Dirac cone. They are semiconducting material with a bulk gap of about 300 meV and a non perfect stoichiometry, due to Se vacancies and Te defects, respectively. This makes them n-type and p-type degenerate semiconducting materials, respectively. By transport measurements it has been found that Ca doping renders Bi_2Se_3 p-type, leading the Fermi energy level from the conduction band toward the valence band. Hence, $Bi_{2-x}Ca_xSe_3$ shows a typical insulating resistivity, improving its TI characteristic. Moreover, it has been proved that compensating Bi_2Se_3 and Bi_2Te_3, new more insulating alloys can be grown, such as Bi_2Se_2Te and Bi_2Te_2Se. This latter has a low temperature resistivity of 5 Ω cm.

In the present work we have studied, by means of infrared spectroscopy, the low energy optical conductivity of those compounds, in order to identify the extrinsic charge contribution of the bulk and to separate it from the intrinsic contribution of the SS carriers.

In particular we have measured the reflectivity of four crystals (Bi_2Se_3, $Bi_{1.9998}Ca_{0.0002}Se_3$, Bi_2Se_2Te and Bi_2Te_2Se) from the Sub-THz to the visible frequency range at different temperatures. Once the optical conductivity has been extracted by the Kramers-Kronig transformations, we have separated the phononic contribution to the electronic one through a Drude-Lorentz-Fano fit. The phonon absorption has been studied by a Fano analysis, showing its interaction with a far-infrared electronic band, due to the transitions of the bulk impurity states, which represent the greatest contribution to the low-energy optical conductivity even in the most compensated single crystals here measured.

Given that the topological electrodynamics at the surface of TI single crystals appeared to be masked by extrinsic conductivity, we further investigate the optical properties of TI by using thin films. We have measured the FIR transmittance of Bi_2Se_3 thin films on sapphire substrate of two different thickness (60 and 120 nm). Their conductance reveals a nearly independence of the free carrier contribution (Drude term) on the film thickness. This suggests that the free-carrier contribution comes from surface states, providing evidence for a 2D electron gas. Comparing the optical conductivity of films ad single crystals, we have deduced that TI thin films are the best candidates to probe the topological surface state by optics. The study of the carrier density, as determined by the optical spectral weight, both of the crystals and films, proves that 3D charge carrier density in crystals is still higher by a factor of 3 than the one in thin films.

In order to further study the 2DEG in TI films we have fabricated by Electron Beam Litography (EBL) and Reactive Ion Etching (RIE) patterned samples, which gave us the opportunity to excite Surface Plasmons Polariton (SPPs). They are, indeed, collective modes of surface charge density, that can be excited when the polarization of incident light is perpendicular to the wires of a patterned sample. Hence, we have measured the extinction coefficient of four films with different grating period. What we have observed (in perpendicular light polarization) was a spectral feature showing a Fano line shape, due to the interference (Fano resonance) between one of the bulk phonons and a SPP. Fitting the extinction coefficient to a function, that takes into account the Fano interference, we have extracted the frequency of the bare SPP.

In a 2DEG the frequencies of SPPs have to scale with the square root of the wave vector and such a dispersion law has been verified in our experiment. This definitely shows the 2D character of the carriers in our TI samples.

The thesis is organized as follow.

In the first chapter a review of TIs, both from a theoretical point of view and from an experimental one, is reported. In particular we describe the crystallographic, chemical and transport properties of the samples studied in this thesis.

In the second chapter we describe the experimental apparatus and the technique (Fourier Transform Infrared Spectroscopy, FTIR) that we have used for our measurements. The theory of optical properties and SPPs are also reported. Furthermore, the sample preparation, both for crystals and films, included the patterning process, are described. Then, the analysis procedures including Drude-Lorentz and Fano fits are discussed.

Finally, in the third chapter we show, analyze and discuss all the experimental data and their consequences.

In the last seven years it has been discovered that certain materials, called Topological Insulators (TIs), have exotic metallic states on their surface. These states are characterized by topological properties (see later), that render the electron motions insensitive to scattering by non-magnetic impurities. Such TIs, hence, may provide new ways to bring about novel phases and quasi-particles in condensed matter, hopefully finding applications in technology, for instance in spintronics or quantum computing.

TIs are insulating in the bulk and conducting on their surface and behave like a plastic cable covered with a layer of metal, except that the material is actually the same throughout. Moreover, the conducting electrons arrange themselves into spin-up and spin-down states, traveling in opposite directions. Unusually, TIs have been theoretically predicted by Kane and Mele in 2005 [1] before being discovered experimentally. Kane and Mele showed how particular surface states appear in three-dimensional (3D) systems allowing for a metallic conduction in otherwise 3D insulating materials. When, in 2007, the TIs were discovered experimentally, they drew the attention of the condensed-matter-physics community, even if a related phenomenon, the Quantum Hall Effect (QHE), had already been found in the early 1980s in 2D systems at very low temperatures. The 3D TIs, moreover, are fairly standard bulk semiconductors and their topological properties survive to high temperature, thus leading to potentially applications.

The quantum-mechanical behavior of electrons in materials entails many important phenomena in condensed matter physics. The conventional insulating state (see Fig. 1.1a) occurs when an energy gap is opened, separating the occupied electronic states from the empty ones and it is due to the quantization of the energy of atomic orbitals. Another more exotic insulating state is the Quantum Hall phase, which provides a conductance precisely quantized in units of fundamental physical constants ($\sigma_{xy} = -\frac{ne}{B}$), when the material is at very low-temperature. Such a state occurs when electrons confined to a 2D interface between two semiconductors experience a magnetic field. These electrons are forced to move into a circular orbit (see Fig. 1.1b), corresponding to an atomic orbital with quantized energy. This leads to an energy gap separating the occupied and empty states, just like in an ordinary insulator. At the boundary of the system, however, the electrons undergo a different kind of motion, because the circular orbits can bounce off the edge, leading to open "skipping orbits", as shown in Fig. 1.1b. In quantum theory these skipping orbits lead to electronic states that propagate along the edge in one direction having not quantized energies. Therefore these states can conduct. The Quantum Hall state can be considered the first topological insulator, as it exhibits a " topological order", which is characterized, rather than by a breaking of symmetry (like for example in a superconductor), by a different topological behavior of the quantum states of electrons (i.e. of their wavefunctions) (see Sect. 1.2.4). Furthermore, the gapless edge state exhibits a one-way flow of electrons, thus the transport is "dissipationless ", i.e protected from impurity scattering.

Unlike the quantum Hall effect, which occurs only when a strong magnetic field B is present, topological insulators work in the absence of B. In these materials the role of B is played by the spin-orbit coupling. In atoms with a high atomic number, such as mercury and bismuth, the spin-orbit coupling is very strong. Hence, electrons traveling through materials composed of such atoms feel a strong spin- and momentum-dependent force, that resembles a magnetic field, the direction of which changes when the spin changes.

The simplest 2D topological insulator is represented by the so called quantum spin Hall state (see Fig. 1.1c). First predicted in 2005, the quantum spin Hall effect occurs when the spin-up and the spin-down electrons, which feel equal and opposite spin-orbit "magnetic fields", are each in quantum Hall states. Then, there are edge states in which the spin-up and the spin-down electrons propagate in opposite directions: the Hall conductance is zero, because the two motions cancel each other. However, such edge states can conduct, since they form a one-dimensional conductor. Like the quantum Hall edge states, the quantum spin Hall edge states are protected from backscattering. Actually, in the latter case the protection arises not only from backscattering but also from the Time Reversal (TR) symmetry: if in the quantum Hall effect the magnetic field breaks the TR of the edge state, in the quantum Spin Hall effect TR switches both the direction of propagation and the spin direction, interchanging the two counter-propagating modes. TR plays a fundamental role to guarantee the stability of the topological states.

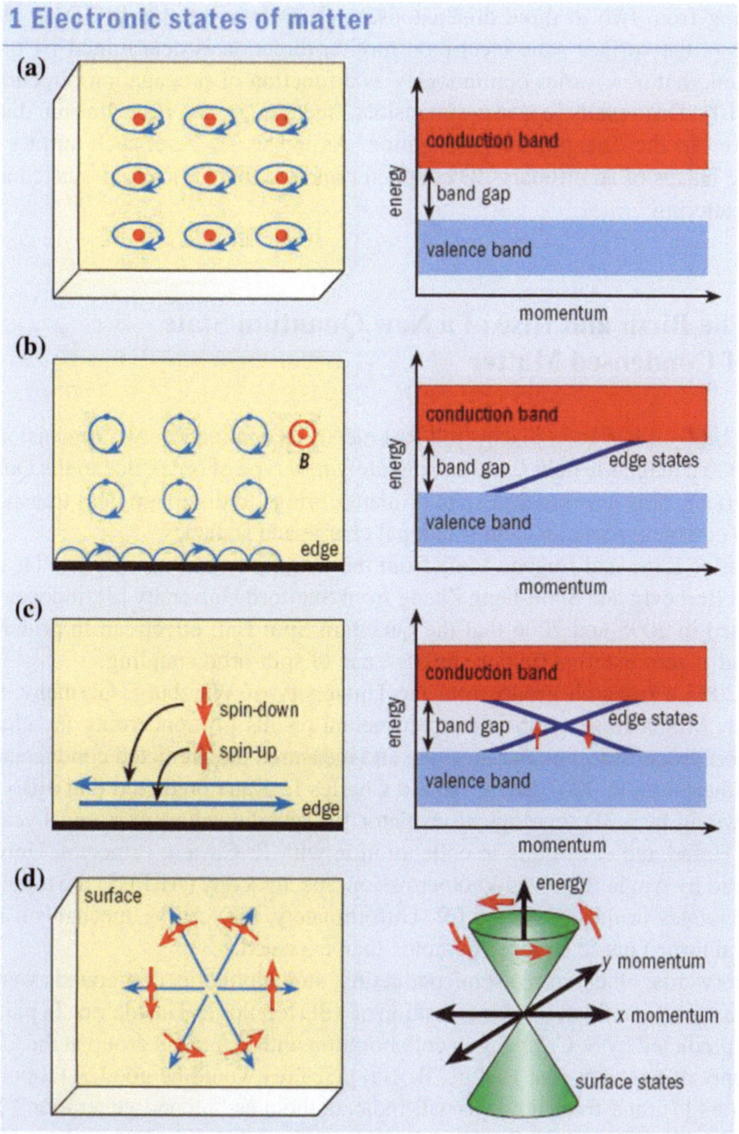

Fig. 1.1 The insulating state characterized by an energy gap separating the occupied and empty states, which is a consequence of the quantization of the energy of atomic orbitals (**a**). In the Quantum Hall Effect the circular motion of electrons in a magnetic field is interrupted by the sample boundary: at the edge the electrons describe "skipping orbits", leading to a perfect conduction in one direction along the edge (**b**). The edge of the Quantum Spin Hall Effect (QSHE) or 2D TI contains left-moving and right-moving modes, that have opposite spin and are related by time reversal symmetry. This edge can also be viewed as half of a quantum wire, which would have spin-up and spin-down electrons propagating in both directions (**c**). The surface of a 3D TI supports electronic motion in any direction along the surface, but the direction of the electron's motion uniquely determines its spin direction and *viceversa*. The 2D energy-momentum relation has a "Dirac cone" structure similar to that in graphene (**d**) [15]

Going from two to three dimensions, one can imagine that in 3D topological insulators the surface state electronic motion direction is determined by the spin direction, that now varies continuously as a function of propagation direction (see Fig. 1.1d). This results in a two-dimensional metallic state where the spin direction is locked to the direction of propagation. As in the 2D case, such surface states are like halves of an ordinary 2D conductor and are topologically protected against backscattering.

1.2 The Birth and Rise of a New Quantum State of Condensed Matter

In the 1980s it has been discovered that electrons confined in two dimensions and subject to a magnetic field show a completely new type of order, due to the Quantum Hall Effect. This new ordered state of matter brings to dissipationless transport but also to emergent particles with fractional charge and statistics.

Charles Kane and Eugene Mele from the University of Pennsylvania [1, 2] and Andrei Bernevig and Shou-heng Zhang from Stanford University [3], independently proposed in 2005 and 2006 that the Quantum Spin Hall effect can in principle be realized at zero external field in the presence of spin-orbit coupling.

In 2007 a research group from the University of Würzburg, Germany, led by Laurens Molenkamp, published experimental results [4] concerning the electrical transport properties of quantum wells, and measured the predicted conductance.

Furthermore, in 2007 Liang Fu and Charles L. Kane predicted that a $Bi_{1-x}Sb_x$ alloy would be a 3D topological insulator for specific values of x and 1 year later Zahid Hasan and coworkers in collaboration with R. Cava at Princeton University observed by Angle Resolved Photoemission Spectroscopy (ARPES) the topological surface states in those systems [6]. Unfortunately, the leading mechanism in that material turned out to be more complex than expected.

Afterwards, other simple semiconducting, stoichiometric, compounds were proposed as good candidates for a realization of a 3D topological insulators. In particular it was predicted by S. C. Zhang in collaboration with Z. Fang's group at the Chinese Academy of Sciences that Bi_2Te_3, Bi_2Se_3 , Sb_2Te_3 would be good 3D topological insulators [7], and from now we will indicate them as "second generation" TIs. In particular, Bi_2Te_3 and Bi_2Se_3 have been investigated at Princeton, where Hasan's group observed in ARPES experiments a single Dirac cone.

A relatively large bulk gap (about 300 meV), the presence of a single Dirac cone (see next section) and the persistence of the topological behavior at room temperature make those materials the best candidates to investigate their electronic properties by transport and optical measurements, as it has been done in the present work, in order to elucidate the contribution to the optical conductivity of the surface carriers in those systems.

1.2.1 The 2D Electron Gas

In the 70s, one of the most important developments in semiconductors, both from the point of view of physics and for the purpose of device developments, has been the achievement of structures in which the electronic behavior is essentially two-dimensional (2D), forming the so-called *two Dimensional Electron Gas* (2DEG). This means that the carriers are trapped in a potential, such that their motion in one direction is restricted and quantized, leaving only a two-dimensional momentum k to characterize the motion in a plane perpendicular to the trapping potential. The most important systems where such a 2D behavior has been studied are the MOS (Metal Oxide Semiconductor) structures, quantum wells and superlattices. More recently, quantization has been achieved in one-dimension (quantum wires) and "zero"-dimensions (quantum dots), but also in three-dimensional topological insulators, that are the subject of this work.

The two-dimensional system, for instance in a semiconductor heterostructure, is described on the basis of a 3D crystal in the effective mass approximation. Taking the confining potential along the z direction, the electrons are free to move in the xy plane. Assuming a single electron scheme and that the effective mass be the same in the different layers of the heterostructure, the eigenfunction for the Schrödinger equation can be separated, i.e

$$\Psi(x, y, z) = e^{ik_x x} e^{ik_y y} u_n(z) \tag{1.1}$$

and the energy eigenvalues are

$$E_n(k_x, k_y) = \frac{\hbar^2 k_x^2}{2m} + \frac{\hbar^2 k_y^2}{2m} + \epsilon_n = \frac{\hbar^2 \mathbf{k}^2}{2m} + \epsilon_n \tag{1.2}$$

where $\mathbf{k} = (k_x, k_y)$ is the two-dimensional wave vector, n is the quantum number of the confinement potential and m is the effective mass . The function $u_n(z)$ and the ϵ_n are the solution of the z-part of the Schrödinger equation. The density of states per unit area $D(E)$ for one subband of a system with degeneracy g_s is constant and equal to $g_s m/2\pi\hbar^2$, when the energy is greater than ϵ_n. The density of states (DOS) is a step-like function when more subbands are considered. The electron density per unit area is then given by

$$n_{2D} = \int_{-\infty}^{\infty} D(E) f(E, E_F) dE \tag{1.3}$$

where $f(E, E_F)$ is the Fermi distribution with Fermi energy E_F. In the low temperature regime Eq. 1.3 becomes

$$n_{2D} = \frac{m}{\pi\hbar^2} \sum_j (E_F - \epsilon_j)\Theta(E_F - \epsilon_j) \tag{1.4}$$

where the sum runs over the subbands with minimum at ϵ_j.

Now, let us consider the effect of a magnetic field perpendicular to the 2DEG, i.e parallel to the confinement direction z. The magnetic field is included in the Hamiltonian through the vector potential and in the Schrödinger equation through the minimal substitution. Choosing the Landau gauge for the vector potential **A** ($\mathbf{A} = (0, Bx, 0)$), the Hamiltonian becomes:

$$H = \frac{\hbar^2}{2m}k_x^2 + \frac{\hbar^2}{2m}(k_y^2 + eBx)^2 + \frac{\hbar^2}{2m}k_z^2 + V(z) \tag{1.5}$$

from which one can obtain the discrete eigenvalue ϵ_n separating the z-part. Being the Hamiltonian independent on y, the eigenfunction is the product of a plane wave e^{iky} and a wave function $\upsilon(x)$. After rearranging the terms and substituting the y-part of the wave function, one can finally find that the x-part of the Hamiltonian is that one of an harmonic oscillator, whose Schrödinger equation is

$$\left[\frac{\hbar^2}{2m}\frac{\partial^2}{\partial x^2} + \frac{1}{2}m\omega_c^2\left(x - \frac{\hbar k}{eB}\right)^2\right]\upsilon(x) = E\upsilon(x) \tag{1.6}$$

where the angular frequency

$$\omega_c = \frac{eB}{m} \tag{1.7}$$

is the so called cyclotron frequency (about $50\,\mathrm{cm}^{-1}$ for $B = 1$ T [30]). Hence, the magnetic field acts as a parabolic confinement potential in the x direction and electrons move in orbits, whose center is

$$x_k = \frac{\hbar k}{eB} = l_B^2 k \tag{1.8}$$

where l_B is the length scale equal to $\sqrt{\hbar/eB}$.

The discrete energy eigenvalues are

$$E_i = \hbar\omega_c\left(i + \frac{1}{2}\right), i = 0, 1, 2... \tag{1.9}$$

known as "Landau levels". The eigen functions are

$$\upsilon(x) = \upsilon_{nk}(x) \propto H_{n-1}\left(\frac{x - x_k}{l_B}\right)e^{-\frac{(x-x_k)^2}{2l_B^2}} \tag{1.10}$$

where $H_n(x)$ is the n-th Hermite polynomial [16].

The effect of the Landau states on the electronic properties of the 2DEG will be described in Sect. 1.2.3.

1.2.2 The Quantum Hall Effect

In 1980, about a century after the discovery of the Hall effect, K. von Klitzing, studying the Hall effect in a 2DEG of a Silicon MOSFET (Metal-Oxide-Semiconductor Field-Effect-Transistor) under a strong magnetic field ($3 \div 10$ T) and at a temperature of about 1.5 K [17], obtained an anomalous result. Rather than the classical Hall resistivity (resistivity in the xy plane), linearly depending on the magnetic field B

$$\rho_{xy} = -\frac{B}{nec} \tag{1.11}$$

where n is the number of charge carriers, e the electronic charge and c the light velocity, whose related conductivity in the xy plane is

$$\sigma_{xy} = -\frac{ne}{B} \tag{1.12}$$

he found

$$\rho_{xy} = \frac{1}{i}\frac{h}{e^2} \tag{1.13}$$

where i is an integer (see Fig. 1.2) and the related conductivity is

$$\sigma_{xy} = \frac{ie^2}{h} \tag{1.14}$$

The Hall resistivity is then quantized and the quantization is universal, independent on the semiconductor structure and with an accuracy of one part/10^9. As shown in

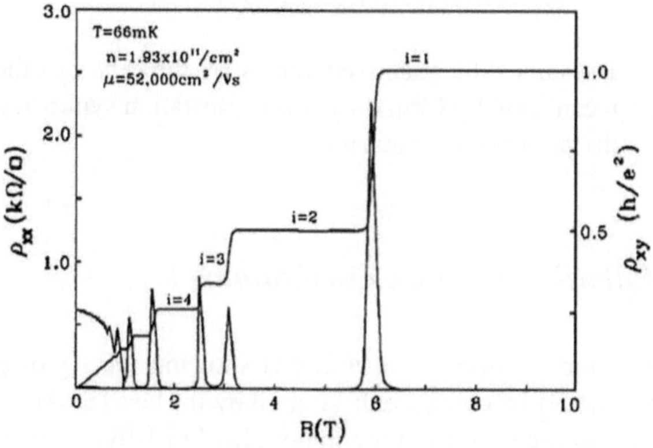

Fig. 1.2 The resistivity along xx and xy planes, ρ_{xx} and ρ_{xy} respectively, versus the magnetic field for the Integer Quantum Hall Effect (IQHE) states. The former quantity (*left scale*) is zero when the latter one (*right scale*) has a plateau [8]

Fig. 1.2 another important characteristic of the so called Integer Quantum Hall Effect (IQHE) is that the longitudinal resistivity ρ_{xx} is zero in correspondence of the ρ_{xy} plateau, while shows peaks when ρ_{xy} is not defined. In 1985 the Nobel Prize in Physics was awarded to von Klitzing.

The finite dimension of the sample in which 2DEG occurs provides some consequences. As we have seen in Sect. 1.2.1, an electron in a magnetic field has quantized wave number i.e $k_{x,y} = 2\pi i/L_{x,y}$ with $L_{x,y}$ the longitudinal dimensions of the sample and i an integer, such that

$$0 < i < \frac{L_x L_y}{2\pi l^2} \tag{1.15}$$

Here l is the magnetic length equal to $\sqrt{\hbar c/eB}$. This maximum value for i corresponds to the degeneracy of the 2DEG (N_{DEG}) and is given by the ratio between the area of the sample and that associated with the quantum flux, equal to $\Phi_0 = 2\pi l^2 B = \frac{hc}{e}$. Hence, N_{DEG} is equal to Ψ/Ψ_0, that is to the number of quantum elementary fluxes which flow through the sample. For every Landau level there are N_{DEG} electrons. If N is the total number of electrons, one can define the *filling factor* as

$$\nu = \frac{N}{N_{DEG}} \frac{hc}{eB} = n\frac{hc}{eB} \tag{1.16}$$

where here n is the 2D electronic density. Since $\nu < 1$, an electron added to the system will go into the first Landau level, not yet filled. The number of filled Landau levels is ν. The energy necessary to add an electron to the system in another new empty level is $\hbar\omega_c$.

Therefore, the magnetic field can be written as

$$B = \frac{nhc}{\nu e} \tag{1.17}$$

If $\nu = 1$, Eq. 1.11 assumes the quantized values for the center of the plateau 1.13. Such these plateau can actually form only if the translation symmetry is broken, for instance because the presence of impurities.

1.2.3 The Shubnikov de Haas Oscillations

In a magnetotransport experiment of a 2DEG showing the IQHE phenomenon a prominent effect should be considered: Shubnikov-deHaas (SdH) oscillations.This effect corresponds to oscillations in the longitudinal resistivity (or conductivity) as a function of a magnetic field. The effect involves electrons which move in a plane perpendicular to the applied field. The SdH effect is a standard test to verify the actual presence of a 2DEG in a sample.

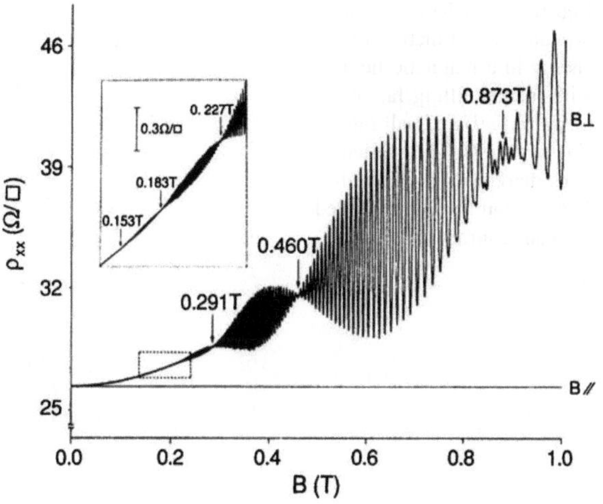

Fig. 1.3 Examples of Shubnikov deHaas oscillations in a heterostructure of InGaAs [19]. Beating nodes are visible, related to a two spin subband occupancy; the line (marked with $B_{||}$) corresponding to the magnetic field parallel to the 2DEG is also reported: no effect is seen in this case

One can also recognize the presence of spin-orbit interaction by beatings in the oscillations. The main feature, i.e. the oscillations, of SdH are described in this section. Other details will be discussed in connection with the experimental results in Sect. 1.3.2. Figure 1.3 shows an example of experimental magnetoresistivity curves for a heterostructure of InGaAs. Different features of the SdH effect can be observed: oscillatory trend, damping, temperature behavior and effect of spin splitting. The oscillations result from the emptying of the Landau levels as a function of increasing magnetic field. The energy of the n-th Landau level (see Eq. 1.9) is

$$E_i(B) = \hbar\omega_c\left(i + \frac{1}{2}\right) = \frac{\hbar e}{m}B\left(i + \frac{1}{2}\right), \quad \Delta E(B) = \frac{\hbar e}{m}B \tag{1.18}$$

with i integer. As seen in the previous section, the Landau levels degeneracy is a function of the magnetic field; indeed, the density of states of the 2DEG at zero field rearranges in a series of discrete levels, whose spacing increases with the magnetic field. Using the densities per unit area, the degeneracy of the Landau level per unit area, which appears in Eq. 1.16, can be calculated from the 2DEG DOS at zero field, i.e $D_0 = g_s m/2\pi\hbar^2$ (where $g_s = 2$ counts the two spin states in the same Landau level), as

$$N_{DEG} = D_0\Delta E(B) = g_s\frac{e}{h}B \tag{1.19}$$

As a function of the magnetic field, if the total number of electrons is fixed, the Fermi energy is pinned to the highest occupied Landau level and it switches from one level to the other at critical values of the field, corresponding to integer values of the filling factor. When the number of filled Landau levels is an integer n, the Fermi level E_F lies between the i-th and $i+1$-level, as shown in Fig. 1.4. It is worth noting that in that Figure the Landau levels have a finite width, to ease the understanding of the

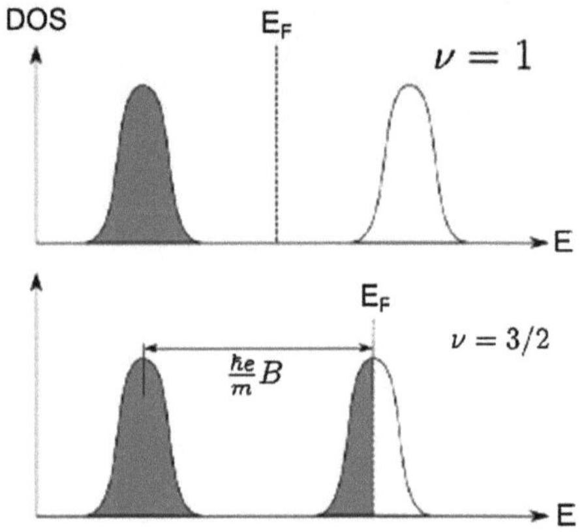

Fig. 1.4 Sketch of the density of states as a function of energy in a magnetic field, with integer filling factor (*upper panel*) and half integer (*lower panel*). The vertical lines mark the Fermi level. The Landau levels are plotted with an arbitrary lineshape

picture, while in this discussion the DOS is assumed to be a series of delta function. The DOS at E_F is then 0; when E_F lies inside a level, the DOS will be different from 0 and have a maximum for half integer values of the filling factor.

Since the conductivity is proportional to the DOS at E_F, it will be zero [1] when E_F lies between two levels and has a maximum when the DOS has a maximum.

The periodicity of the maxima (or minima) in the conductivity as a function of the magnetic field is given by the periodicity of the filling factor: in particular, when $\nu = i$ is fullfilled at

$$\frac{1}{B_i} = \frac{e}{h}\frac{g_s}{N_{DEG}} i \tag{1.20}$$

showing how the minima are evenly spaced in the inverse magnetic field, that is

$$\frac{1}{B_i + 1} - \frac{1}{B_i} = \frac{e}{h}\frac{g_s}{N_{DEG}} \tag{1.21}$$

Then, the 2DEG density can be calculated from the position of the extrema or, easily, from the oscillation frequency f, given by

$$f = \frac{h}{g_s e} N_{DEG} \quad \Leftrightarrow \quad N_{DEG} = g_s \frac{e}{h} f \tag{1.22}$$

When two parallel conductive channels with different carrier density are present in the sample, the sum of the two oscillations causes a modulation in the magnetoresistance (beatings). The two frequencies f_1 and f_2 can be extracted from a Fourier spectrum or from the minima of the envelope of the oscillations.

[1] Actually in real samples conductivity is not zero, because the DOS is finite between the Landau levels.

In the case of a spin-split conduction band, interesting for this work, there are two conductive channels with very similar frequency and amplitude; hence, a beating pattern with node is expected.

1.2.4 Why Topological?

What is the difference between a quantum Hall state and an ordinary insulator? The answer, first proposed in 1982 by Thouless, Kohomoto, Nightingale and den Nijis (TKNN), is a matter of topology. Topology is an area of mathematics concerned with properties that are preserved under continuous deformations of objects, such as deformations that involve stretching, but no tearing or gluing. A 2D band structure consists of a mapping from the crystal momentum $\mathbf{k}$ (defined on a torus) to the Bloch Hamiltonian $H(k)$. Band structures with a finite energy gap, which separates the occupied and the empty bands, can be classified topologically by considering the equivalence classes of $H(k)$, that can be continuously deformed into another one without closing the energy gap. Equivalently, if one considers the set of N occupied Bloch wavefunctions $\{u_m(\mathbf{k})\}$, defining the many body ground state, this will be unchanged by a rotation $U(N)$ among the occupied states. Hence, $\{u_m(\mathbf{k})\}$ form a $U(N)$ vector bundle over the Brillouin zone torus. The topological classification of those bundles has been well known by mathematicians for 60 years [20]. The integers $\mathbb{Z}$ distinguish the bundles in various *Chern classes*, associated with the so called *Chern number*, given by

$$n = \frac{1}{2\pi}\int d^2\mathbf{k}\mathcal{F} \tag{1.23}$$

Here, $\mathcal{F} = \nabla \times \mathcal{A}$ is the *Berry's curvature*, where $\mathcal{A} = i\sum_{m=1}^{N}\langle u_m|\nabla_k|u_m\rangle$ is the Berry's connection. Equation 1.23 can be interpreted as the *Berry's phase* acquired by the valence band states when they are adiabatically transported around the perimeter of the Brillouin zone [21]. It is similar to a winding number, that is the integer number of a closed curve in the plane around a given point, representing the total number of times that the curve travels counterclockwise around the point (it depends on the orientation of the curve and is negative if the curve travels around the point clockwise).

TKNN showed that the Hall conductivity, computed using the Kubo formula, has exactly the same form as Eq. 1.14, so that n is identical to i in Eq. 1.14. Furthermore, the Chern number is a topological invariant in the sense that it can not change when the Hamiltonian smoothly varies, for instance because of a weak perturbation.

The meaning of Eq. 1.23 can be better understood by an analogy, considering a simple map from two dimensions into three dimensions, namely from curves to surfaces. Two-dimensional surfaces can be topologically classified by their genus, g, which counts the number of holes: for examples, a sphere has $g = 0$, while

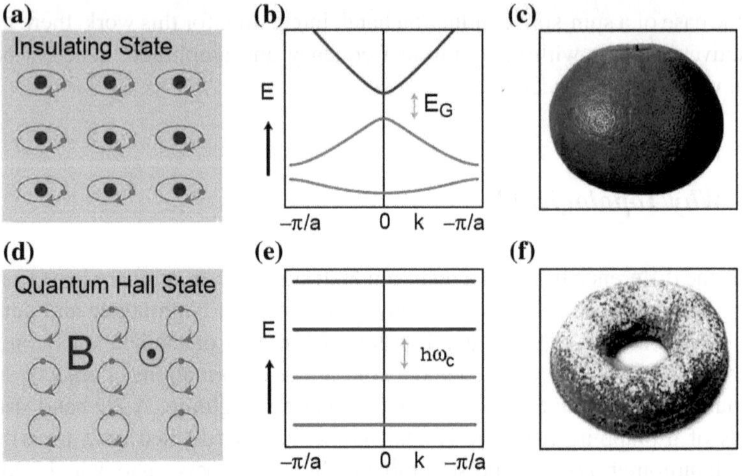

Fig. 1.5 The insulating state (**a**, **b**, **c**), where **a** depicts an atomic insulator and **b** its simple model band structure. The quantum Hall state (**d**, **e**), where **d** depicts the cyclotron motion of electrons and **e** the Landau levels in the energy distribution, which may be viewed as a band structure. **c** and **f** show, in analogy, two surfaces which differ in their genus g: for the sphere $g = 0$, while for the donut (torus) $g = 1$ [9]

a torus has $g = 1$ (see Fig. 1.5). A mathematical theorem due to Gauss and Bonnet [20] asserts that the integral of the Gaussian curvature over a closed surface is a quantized topological invariant and its value is related to g: the Chern number, which is the integral of a curvature, is a similar kind of invariant. Therefore, a topological insulator can be explained as a material constructed by an insulating state in the bulk and a quantum Hall state related to the surface state at the edge. These two states are associated with different topological phases: one cannot cross the material under the same continuous deformation without changing the topological invariant. That explains the name chosen for this new class of materials in condensed matter.

An other easy way to understand that underlying phenomenon is to consider an intuitive illustration, showing why in a TI the metallic surface exists. In Fig. 1.6 a trefoil knot is used to represent a topological insulator while a closed loop to represent an ordinary one. Those two surfaces have different topological invariants and thus neither can be deformed to become the other, no matter how the string is stretched or twisted, without being cut. Nevertheless, those invariants must change in crossing the interface between topological and ordinary insulators, so by contradiction the surface can not remain insulating, which would be analogous to cutting the knot.

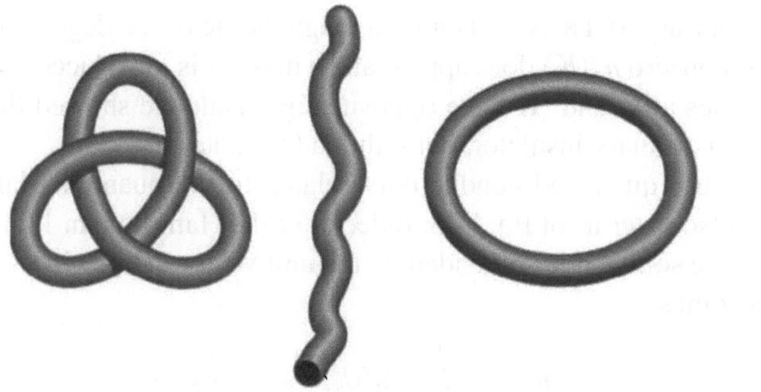

Fig. 1.6 An illustration of topological change and the resultant surface state. The trefoil knot (*left*) and the simple loop (*right*) represent different insulating materials: the knot is a topological insulator and the loop is an ordinary insulator. Because there is no continuous deformation by which one can be converted into the other, there must be a surface where the string is cut, shown as a string with open ends (*centre*), to pass between the two knots; more formally, the topological invariants cannot remain defined. If the topological invariants are always defined for an insulator, then the surface must be metallic [10]

1.2.5 The Edge State and the Dirac Fermions

A simple example of QHE in the band theory is provided by a model introduced by Haldane [11]. As in graphene, in TIs the conduction band and the valence band touch each other at two distinct points in the Brillouin zone. Near those points the electronic dispersion resembles the linear dispersion of massless relativistic particles, described by the Dirac equation [12]. The simplest description of graphene employs a two band model for the p_z orbitals on the two equivalent atoms in the unit cell of graphene's honeycomb lattice. The Bloch Hamiltonian is then the 2×2 matrix

$$H(\mathbf{k}) = \mathbf{h}(\mathbf{k}) \cdot \sigma \tag{1.24}$$

where $\sigma = (\sigma_x, \sigma_y, \sigma_z)$ are Pauli matrices and $\mathbf{h}(\mathbf{k}) = (h_x(\mathbf{k}), h_y(\mathbf{k}), 0)$, where $h_z(\mathbf{k}) = 0$ according to the inversion (P) and TR symmetries. The Dirac points occur because the two components of $\mathbf{h}(\mathbf{k})$ can have solutions in two dimensions. In graphene the Dirac points occur at points $\mathbf{K}$ and $-\mathbf{K}$, whose positions in the Brillouin zone are fixed by graphene's rotational symmetry. For small $\mathbf{q}$ (equal to $\mathbf{k} - \mathbf{K}$), one has $\mathbf{h}(\mathbf{k}) = \hbar v_F \mathbf{q}$ (where v_F is the Fermi velocity) and the Hamiltonian becomes the 2D massless Dirac one:

$$H(\mathbf{k}) = \hbar v_F \mathbf{q} \cdot \sigma \tag{1.25}$$

The degeneracy at the Dirac point is protected by P and TR and can be removed by breaking these symmetries. If P is violated, the two atoms in the unit cell are inequivalent, then the Dirac Hamiltonian becomes massive, with mass $m = h_z(\mathbf{K})$, and the energy dispersion becomes $E(\mathbf{q}) = \pm\sqrt{|\hbar v_F \mathbf{q}|^2 + m^2}$, with an energy gap $2|m|$. TR guarantees that at the Dirac point -$\mathbf{K}$ the mass m' is equal to m.

On the contrary, if TR is broken by a magnetic field, the degeneracy is removed once again: nonzero $h_z(\mathbf{K})$ does appear and a mass m is introduced. Now, P requires that the masses at $\mathbf{K}$ and -$\mathbf{K}$ have opposite sign. Haldane showed that this gapped state is not an ordinary insulator, but rather a QH state.

The non-zero quantized conductivity related to the quantum Hall state can be understood also in terms of Eq. 1.23. Indeed, for the Hamiltonian 1.24 the Berry flux is related to the solid angle subtended by the unit vector $\hat{h}(\mathbf{k}) = \mathbf{h}(\mathbf{k})/|\mathbf{h}(\mathbf{k})|$, so that Eq. 1.23 becomes

$$n = \frac{1}{4} \int d^2\mathbf{k}(\partial_{k_x}\hat{h} \times \partial_{k_y}\hat{h}) \cdot \hat{h} \tag{1.26}$$

and counts the number of times $\hat{h}(\mathbf{k})$ envelops the unit sphere as a function of $\mathbf{k}$. When the masses m and m' are zero (no time reversal breaking), $\hat{h}(\mathbf{k})$ is confined to the equator $h_z = 0$, with one winding around each of the Dirac points (opposite to each other), where $|\mathbf{h}| = 0$. When TR is broken, a small but finite m is introduced and $|\mathbf{h}| \neq 0$ everywhere, so that $\hat{h}(\mathbf{k})$ visits the north or south pole of the sphere (depending on the sign of m). Hence, each Dirac point contributes to the Hall conductivity with $\pm e^2/2h$. In other terms in the insulating state ($m = m'$) the two contributes cancel each other, in the quantum Hall state they add.

It is worth noting that it must be an even number of Dirac points. This fact is also guaranteed by the *fermion doubling theorem*, which states that for a TR invariant system Dirac points must come in pairs. In three dimensions a *single Dirac cone* (with one only Dirac point) has been measured: this fact seems to violate the fermion doubling theorem, except that the partner Dirac point resides on the opposite surface [9].

As seen in the previous section, a quantum Hall state provides at the interface with vacuum an edge state, which is *chiral* in the sense that the electronic motion responsible for that state is in one direction. The edge states are insensitive to disorder because they don't experience backscattering (see next section).

The presence of such one directional edge states is deeply related to the topology of the bulk quantum Hall state. In the previous section we have also introduced the Chern number n: it is 0 for an ordinary insulator (or vacuum) and 1 for a QH state. Passing through one to the other the energy gap has to vanish, because otherwise it is impossible for the topological invariant to change. Therefore, low energy electronic states bound to the region where the gap vanishes have to exist: this is the underlying connection between gapless edge state and topology.

A simple theory of the chiral edge state [13] can be developed using the two-band Dirac model. If one considers an interface where the mass m at one of the Dirac points changes sign as a function of y (the axis perpendicular to the interface), one has that $m(y) > 0$ gives the insulator for $y > 0$ and $m(y) < 0$ gives the QH state for $y < 0$ (with m'>0 fixed). Considering now the massive Dirac Hamiltonian with $\mathbf{q} = -i\overrightarrow{\nabla}$

$$H(\mathbf{q}) = \hbar v_F(-i\overrightarrow{\nabla}) \cdot \sigma + m\sigma_z \tag{1.27}$$

it gives an exact solution

$$\Psi_{q_x}(x, y) \propto e^{iq_x x} e^{-\int_0^y dy' m(y') dy'/v_F} \begin{pmatrix} 1 \\ 1 \end{pmatrix} \quad (1.28)$$

with eigenvalue $E(q_x) = \hbar v_F q_x$. This band of states intersects the Fermi energy E_F with a positive group velocity $dE/dq_x = \hbar v_F$ and defines a right moving chiral edge mode.

The chiral edge states in the QHE can be seen explicitly by solving the Haldane model in a semi-infinite geometry with an edge at $y = 0$. In Fig. 1.7a the energy levels as a function of the momentum k_x along the edge are shown. The energy gap near **K** and -**K** is open between the valence and the conduction band, connected each other by a single band, describing states bound to the edge. By changing the Hamiltonian near the surface, the dispersion of the edge states can be modified. For example, $E(q_x)$ could develop a kink, so that the edge states could intersect the Fermi level three times, twice with a positive group velocity and once with a negative one. But the difference between the number of right and left moving modes $(N_R - N_L)$ can not change and it is determined by the structure of the bulk states: this is the so called *bulk-boundary correspondence*, described by

$$N_R - N_L = \Delta n \quad (1.29)$$

where Δn is the difference in the Chern number across the interface.

In Fig. 1.7b electronic states of a TR invariant 2D insulator as a function of k_x along the edge are shown. Depending on the the details of the Hamiltonian near the edge there may or may not be states bound to the edge inside the gap. If they are, the Kramers theorem requires that they are twofold degenerate at the TR invariant momenta $k_x = 0$ and $k_x = \pi/a$: these two special points are labeled $\Gamma_{a,b}$, away from

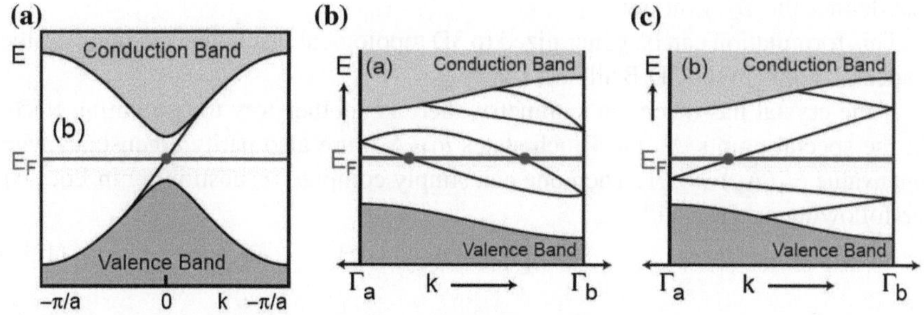

Fig. 1.7 Sketch of a semi-infinite strip described by the Haldane model: a single edge connects the valence band and the conduction band (**a**). Sketch of the same system with electronic dispersion between two boundary Kramers degenerate points $\Gamma_a = 0$ and $\Gamma_b = \pi/a$. In (**b**) the number of surface states crossing the Fermi energy E_F is even, while in (**c**) it is odd. An odd number of crossings leads to topologically protected metallic boundary states [9]

which a spin orbit interaction will split the degeneracy. As one can see in in Fig. 1.7b there are two ways by which the states at Γ_a and Γ_b can connect and there are an even number of times (red points) that those two bands intersect E_F. In this case the edge states can be eliminated by pushing all of the bound states out of the gap. On the contrary, in Fig. 1.7c (linear dispersion) the edge states cannot be eliminated: there will be always one point where the bound states intersect the Fermi level (red point). Therefore, in the latter case, the bands intersect E_F an odd number of times. The occurrence of the former or the latter case depends only on the topological class of the bulk band structure. One can relate the number of Kramers pairs N_K of edge modes intersecting E_F (since the Kramers points are the opposite intersections of E_F at k_x) to the change of the so called $\mathbb{Z}_2$ topological invariants ν across the interface (see also next section), that is

$$N_K = \Delta\nu \text{mod} 2 \tag{1.30}$$

Therefore, a 2D topological insulator ((c) in Fig. 1.7) has a topologically protected edge state.

These considerations can be extended to a 3D topological insulator, as described in the next section.

There are several mathematical formulations of the $\mathbb{Z}_2$ invariant ν. One of those [14] uses the unitary matrix $w_{mn}(\mathbf{k}) = \langle u_m(\mathbf{k})|\Theta|u_n(-\mathbf{k})\rangle$ where $\{u_m(\mathbf{k})\}$ are the occupied Bloch functions and Θ is an antiunitary operator ($\Theta^2 = -1$ and $w^T(\mathbf{k}) = -w(-\mathbf{k})$). In the bulk 2D Brillouin zone there are four special points Λ_a where $\mathbf{k}$ and $-\mathbf{k}$ coincide, so $w(\Lambda_a)$ is antisymmetric. Considering $\delta_a = Pf[w(\Lambda_a)]/\sqrt{Det[w(\Lambda_a)]} = \pm 1$[2] and choosing $|u_m(\mathbf{k})\rangle$ continuously throughout the Brillouin zone, one has

$$(-1)^\nu = \prod_{a=1}^{4} \delta_a \tag{1.31}$$

that defines the $\mathbb{Z}_2$ invariant.

This formulation can be generalized to 3D topological insulators and involves the 8 special points in the 3D Brillouin zone.

If the crystal has inversion symmetry, there is another way to computing ν [5]. At the special points Λ_a the Bloch states $u_m(\Lambda_a)$ are also parity eigenstates with eigenvalue $\xi_m(\Lambda_a) = \pm 1$. Then one can simply compute ν substituting in Eq. 1.31 the following expression

$$\delta_a = \prod_m \xi_m(\Lambda_a) = \pm 1 \tag{1.32}$$

where the product is over the Kramers pair of occupied bands.

Equations 1.31 and 1.32 are useful to identify the topological insulators from band structure calculations.

[2] Here Pf is the Pfaffian of the matrix, that is the polynomial whose square is the determinant of an antisymmetric matrix.

1.2.6 The Spin-Orbit Coupling and the Spin Quantum Hall Effect

We remarked previously that Spin-Orbit Coupling (SOC) has a fundamental role in TIs. Indeed, SOC generates a magnetic field which replaces the external field providing the Quantum Hall Effect (QHE), giving rise to the so called Quantum Spin Hall Effect (QSHE).

SOC is related to the motion of an electron in an electric field. In the rest frame of the electron, the electric field **E** is seen as a magnetic field, which acts on the magnetic dipole of the electron (due to the spin). This effective magnetic field, in SI units, can be written as

$$\mathbf{B} = -\left(\frac{\mathbf{v}}{c^2} \times \mathbf{E}\right) \tag{1.33}$$

where **v** is the electron velocity and c the light velocity.

From classical SOC Hamiltonian $H = -\mu \cdot \mathbf{B}$, one can obtain in the Hamiltonian for a given electric potential $V(r)$:

$$H_{SO} = \frac{e\hbar}{4m^2c^2}\mathbf{p} \cdot (\sigma \times \nabla V(r)) \tag{1.34}$$

where m is the electron mass and σ is the vector built up with the Pauli matrices.

For a central potential one can rewrite $\nabla V(r) = (dV/dr)(\mathbf{r}/r)$ and, rearranging the triple product, one finds

$$H_{SO} \propto (\mathbf{r} \times \mathbf{p}) \cdot \sigma = \mathbf{L} \cdot \mathbf{S} \tag{1.35}$$

with **L** and **S** the orbital angular momentum and the spin, respectively.

In solids, the description of spin-orbit interaction is complicated by the details of the crystal lattice. Qualitatively, if the electronic states are represented in an atomic-orbital picture, one can say that the p-like (L = 1) bands are affected by the spin-orbit, while the s-like (L = 0) are not. This "atomic"part of the SOC is usually taken into account in band structure calculations and is found to affect mainly the valence band. Valence bands of semiconductors are (at the zone center) p-like states. The 3-fold degeneracy is removed by SOC and the $j = 3/2$ and $j = 1/2$ states are split (j is the total angular momentum $\mathbf{J} = \mathbf{L} + \mathbf{S}$). In analogy with the atomic case, the SOC affects the "orbital motion"of electrons, but retains the J_z degeneracy.

Although spin-orbit coupling has not the symmetry required to induce the quantum Hall effect (i.e it does not break TR symmetry as an applied magnetic field does), the simplified models introduced in 2003 can lead to a quantum spin Hall effect (QSHE), in which electrons with opposite spin angular momentum (commonly called spin-up and spin-down) move in opposite directions, for instance around the edge of a droplet in the absence of an external magnetic field (see Fig. 1.8) [22]. In a QH state electrons travel at the edge of the semiconductor, with two counterflows of electrons spatially separated into two different straight path located at the top and bottom of the sample. Compared with a 1D system with electrons propagating in both directions, the top

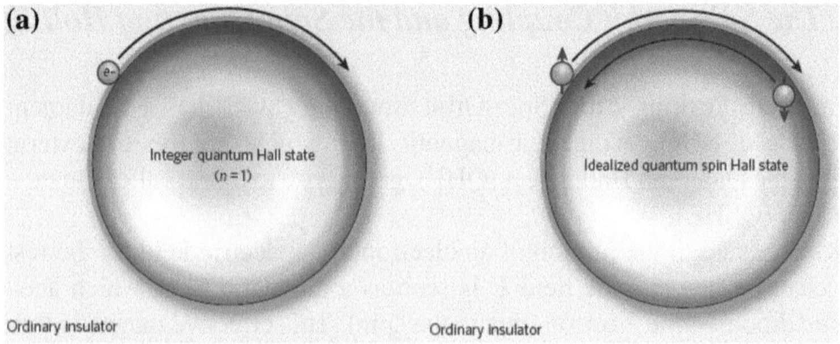

Fig. 1.8 Sketch of an edge of an integer quantum Hall state: the electrons are confined to a 2D insulating droplet with a metallic edge. Along the edge, electrons propagate only in one direction, which is determined by the sign of the applied magnetic field perpendicular to the droplet (**a**). Edge of an idealized quantum spin Hall state (that is, a 2D topological insulator). Along the edge, spin-up electrons move clockwise, whereas spin-down electrons move anticlockwise. Spin-up and spin-down electrons are independent and are in oppositely directed quantum Hall states. An applied electrical field generates a spin current but no charge current (**b**). Each droplet is surrounded by an ordinary insulator [10]

edge of a QH sample contains only half the degrees of freedom. When an electron of the edge state encounters an impurity, it simply takes a deviation and still keeps going in the same direction (see Fig. 1.9 on the left), as there is no way for it to turn back. Such a dissipationless transport process could be very useful for semiconductor devices. Its requirement of a large magnetic field can however be overcame without breaking the two-counterflows electron transport. Indeed, in a real 1D system, the presence of both spin-up and spin-down gives rise to four channels, two forward and two backward (see Fig. 1.9 on the right). Now, the electron path can be split remaining TR invariant, as there is no magnetic field to break the symmetry: one can leave the spin-up forward mover and the spin-down backward mover on the top of the sample and moves the other two channels to the bottom edge. Such a QSH state having a net transport of spin along the top and the bottom of the sample, generates a spin current (but no charge current) just with an electric field applied.

Although a QSH state consists of both backward and forward motions, backscattering by nonmagnetic impurities is forbidden, as it happens for the QHE as shown in Fig. 1.9 on the left.

In analogy with an antireflection coating, for which reflected light from the top and the bottom surfaces interfere with each other destructively, leading to zero reflection, an electron can be reflected by an impurity, causing the interference between different paths. As shown in Fig. 1.10, the electron in a QSH state can take either the clockwise or the counterclockwise turn around the impurity; during that turn the spin rotates by an angle of π or $-\pi$ to the opposite direction. Hence, the two paths, related by TR symmetry, differ by a $\pi - (-\pi) = 2\pi$ full rotation of the spin. According to the quantum mechanics a wavefunction of a spin 1/2 obtains a negative sign upon a full 2π rotation, thus the two backscattering paths always interfere destructively. If the impurity carries a magnetic moment, the TR symmetry is broken and the two

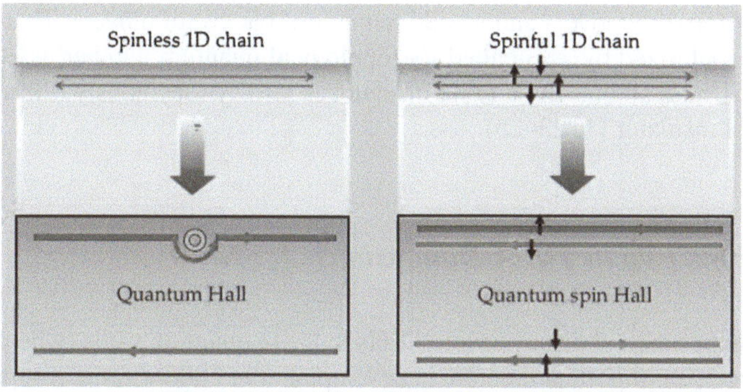

Fig. 1.9 Quantum Hall Effect (*left*) versus Quantum Spin Hall Effect (*right*). A spinless one-dimensional system has both a forward and a backward mover: those two basic degrees of freedom are spatially separated in a QH bar. The upper edge contains only a forward mover and the lower edge has only a backward mover. The states are robust: they will go around an impurity without scattering. On the contrary, a spinful 1D system has four basic channels, which are spatially separated in a QSH bar: the upper edge contains a forward mover with spin-up and a backward mover with spin-down and conversely for the lower edge [23]

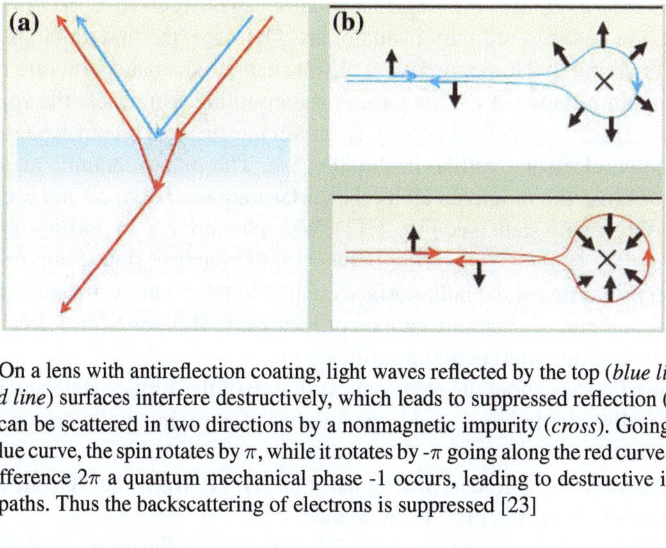

Fig. 1.10 On a lens with antireflection coating, light waves reflected by the top (*blue line*) and the bottom (*red line*) surfaces interfere destructively, which leads to suppressed reflection (**a**). A QSH edge state can be scattered in two directions by a nonmagnetic impurity (*cross*). Going clockwise along the blue curve, the spin rotates by π, while it rotates by -π going along the red curve: associated with the difference 2π a quantum mechanical phase -1 occurs, leading to destructive interference of the two paths. Thus the backscattering of electrons is suppressed [23]

reflected waves no longer interfere destructively: in that sense the robustness against backscattering of the QSH edge state is protected by the TR symmetry.

The physical picture above applies only to the case of single pairs of QSH edge states. If there are two forward motions and two backward motions in the system as, for example, the unseparated 1D system shown in Fig. 1.9 (right), then an electron can be scattered from a forward to a backward moving channel without reversing its spin and without the perfect destructive interference, thus there is dissipation. Consequently, for the QSH state to be robust, the edge states must consist of an odd number of forward movers and an odd number of backward movers. That even-odd effect, characterized by a so-called $\mathbb{Z}_2$ topological quantum number, is at the heart of the QSH state and is why a QSH insulator is also synonymously referred to as a topological insulator [1, 23–25].

1.2.7 From Two to Three Dimensions

After the experimental discovery of 2D TIs in HgTe quantum wells (where the SOC is very large), an important theoretical development in 2006 was the comprehension that even though the quantum Hall state does not generalize to a 3D state, a topological insulator does.

One can form a 3D "weak" topological insulator layering 2D versions (every one with a QH state) but this system results not stable to disorder and its physics is generally similar to that of a 2D state. Indeed, in weak TIs a linear defect of the crystal will always contain a quantum wire like that belongs to the edge state of the QSH state, allowing one to observe the 2D physics in 3D material.

Nevertheless, a "strong" topological insulator can be realized. If in two-dimension the connection between ordinary insulator and QH state (the first topological insulator) is the breaking of TR symmetry, in 3D one can build a band structure respecting TR, once again because of a strong spin-orbit coupling. But in 3D the spin are mix all each other and then there is no way to obtain the strong TI from separate spin-up and spin-down electrons, unlike in the 2D case. The planar metallic state of a 3D TI occurs because the momenta along the surface are well defined and each of them has only a single spin state (see Fig. 1.11). When disorder or impurities are added at the surface, they generate scattering between those surface states, but crucially the topological properties of the bulk-surface connection don't allow the metallic state to vanish, that is it can not become gapped or localized [10] (see Sect. 1.2.6). Actually, this is not true adding magnetic impurities.

Those predictions about the electronic structure of the surface state and its robustness against disorder led to a very large number of experimental work on 3D TIs in the past 4 years.

More theoretically one can distinguish the 2D TIs from the 3D ones by using considerations concerning the TR invariance.

A 3D TI is characterized by four $\mathbb{Z}_2$ topological invariant (see Sect. 1.2.5) $(\nu_0;\nu_1;\nu_2;\nu_3)$ [5]. There are indeed four TR invariant points $\Gamma_{1,2,3,4}$ in the surface Brillouin zone where surface states must be Kramers degenerate and away from which the spin-orbit interaction removes the degeneracy. Those Kramers points form 2D Dirac points in the surface band structure (the center of the Dirac cone as shown in Fig. 1.12c). We have seen in Sect. 1.2.5 how in 2D the surface edge state connects the points Γ_a and Γ_b; now the question is how in 3D the Dirac points connect to each other at different TR invariant points.

The first simple non trivial 3D TI that can be constructed is that formed by layering 2D QSH states: in Fig. 1.12a a sketch of the Fermi surface of such a "weak" 3D TI

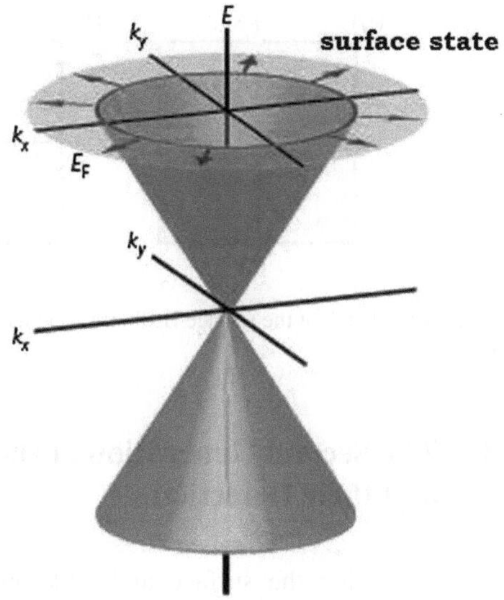

Fig. 1.11 Sketch of theoretical idealization of the electronic structure of a strong TI, showing the rotation of the spin degree of freedom (*red arrows*) as an electron with energy E moves around the Fermi surface (with Fermi energy E_F). Scattering of the surface electrons by non-magnetic disorder will modify the details of the electronic wavefunctions but will not eliminate the metallic surface state [10] (color figure online)

is shown. A single surface band intersects the Fermi surface once between Γ_1 and Γ_2 and once between Γ_3 and Γ_4: an odd intersections between any pair of Γ points leads to the rise of a topological state (as seen in 2D) , that is called weak 3D TI. That weak TI has the invariant number $\nu_0 = 0$. The indexes $(\nu_1 \nu_2 \nu_3)$ can be interpreted as Miller indexes of a crystal and describe the orientation of the layers. Unfortunately those states are not protected against disorder by TR symmetry since they can be localized.

Another distinct phase characterized by $\nu_0 = 1$ is called "strong" 3D TI. Since ν_0 corresponds to an even or odd number of Kramers points enclosed by the Fermi surface, in a strong TI this number is odd, as shown in Fig. 1.12b. The simplest case is having one Dirac point (or equivalently one single Dirac cone), described by the Hamiltonian

$$H_{surface} = -i\hbar v_F \overrightarrow{\sigma} \cdot \overrightarrow{\nabla} \tag{1.36}$$

where $\overrightarrow{\sigma}$ is the spin.

As seen above, one single Dirac cone doesn't violate the fermion doubling theorem, since the partner Dirac point resides on opposite surface.

The surface states of a strong topological insulator form a unique 2D topological metal [5], that is essentially *half* an ordinary metal. Indeed, they are not spin degenerate. Since TR symmetry requires that states at momenta **k** and −**k** have opposite spin, the spin must rotate with **k** around the Fermi surface, as indicated in Fig. 1.12b, that leads also to a finite Berry phase acquired by an electron moving around the Fermi circle. TR symmetry requires that the Berry phase be 0 or π: when the electron circles a Dirac point, its spin rotates by 2π, leading to a π Berry phase.

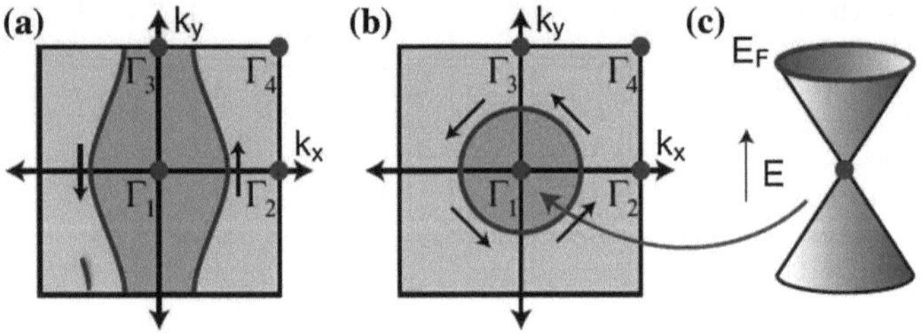

Fig. 1.12 Sketch of the surface Brillouin zone with Fermi circles for a weak TI (**a**) and a strong TI (**b**) [9]

1.3 The Second Generation of the Strong 3D TIs and their Detection

In this section the surface and bulk properties of the strong 3D TIs Bi_2Se_3, $Bi_{2-x}Ca_xSe_3$, Bi_2Se_2Te and Bi_2Te_2Se will be described with experimental results. Those are the samples analyzed in the present work by FTIR (Fourier Transform Infrared Spectroscopy), in order to identify the contribution of bulk carriers and surface carriers as a function of doping and chemical compensation to the optical conductivity.

First the crystal structure and the chemical composition will be illustrated, then the transport data about the resistivity and the detection of the typical Shubnikov deHaas oscillations will be analyzed; the experimental results of Angle Resolved Photoemission Spectroscopy (ARPES) with the related detection of the Dirac cone will be explained; finally, the previous optical results about reflectivity and conductance data of Bi_2Se_3 (for films samples) will be reported.

1.3.1 Crystal Structure and Chemical Composition

Bismuth selenide Bi_2Se_3 is a member of the V_2VI_3 group of materials (being V = Bi, Sb, S and VI = Se, Te, S) and crystallizes in a rhombohedral structure (point group $\bar{3}mD_{3d}$ [29]). In Fig. 1.13 atom layers, known as *quintuple layers* (QL), are oriented perpendicular to the c axis and the covalent bonding within each quintuple layer is much stronger than the weak van der Waals force bonding neighboring layer [30]. The lattice parameters of Bi_2Se_3, together with those of Bi_2Se_2Te and Bi_2Te_2Se, are reported in Table 1.1.

Its semiconducting gap has been measured by transport and optical experiments and turns out to be 0.25–0.35 eV, according to theoretical calculations.

While in the other binary compound, Bi_2Te_3, the chemical similarity of Bi and Te leads to antisite defects as the primary source of carrier doping (p type), in Bi_2Se_3 Bi and Se have a little tendency to mix each other, giving rise to electron doping

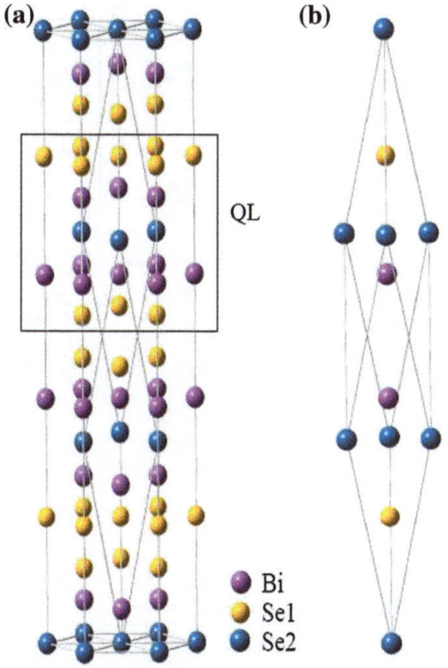

Fig. 1.13 The crystal structure of Bi_2Se_3. The rhombohedral crystal structure of Bi_2Se_3 consists of hexagonal planes of Bi and Se stacked on top of each other along the z direction. A quintuple layer with Se1-Bi-Se2-Bi-Se1 is indicated by the *black square*, where (1) and (2) refer to different lattice positions (**a**). Rhombohedral unit cell of Bi_2Se_3 (**b**) [31]

Table 1.1 Lattice parameters for three TIs samples [28]

	a (Å)	c (Å)
Bi_2Se_3	4.138	28.64
Bi_2Se_2Te	4.218	29.240
Bi_2Te_2Se	4.28	29.86

(n type) by doubly charged Selenium vacancies $V_{Se}^{\bullet\bullet}$ [34] (see the STM image in Fig. 1.17a) . Indeed

$$\mathrm{Se_{Se}} \rightarrow V_{\mathrm{Se}}^{\bullet\bullet} + \mathrm{Se}(g) + 2e'$$

The presence of donors can be compensated for instance by doping with Pb on the Bi site, as Pb has one electron less than Bi. However this substitution does not form a p type material for Bi_2Se_3, probably because Pb is ambipolar [34]. A more ionic substitution on the Bi site can be made by Ca substitution, with the following reaction

$$2\mathrm{Ca} \xrightarrow[\mathrm{Bi_2Se_3}]{} 2\mathrm{Ca}'_{\mathrm{Bi}} + 2h^{\bullet}$$

where Ca substitution for Bi creates a negatively charge defect $\mathrm{Ca}'_{\mathrm{Bi}}$, generating holes $h^{\bullet}$ to compensate the electrons created by the Se vacancies. The crystal structure (for a quintuple layer) of the resulting compound $Bi_{2-x}Ca_xSe_3$ is shown in Fig. 1.14, while the STM topographies are reported in Fig. 1.17 and described later.

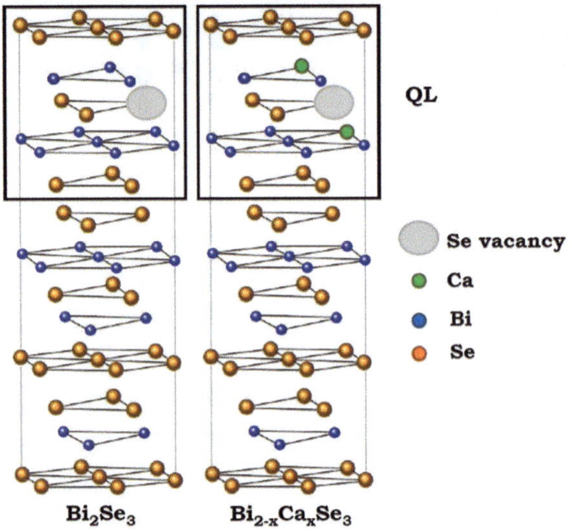

Fig. 1.14 The crystal structure of Bi_2Se_3 with a Se vacancy in the QL and substitution of two Bi with two Ca in $Bi_{2-x}Ca_xSe_3$

The narrow-band semiconducting ternary compounds $Bi_2(Te, Se)_3$ have been studied over 50 years due to their remarkable thermoeletric properties [27].

In binary or ternary semiconductors the small defects density results in large carrier density of $\sim 10^{18-19}$ cm^{-3}, leading the bulk conductance to dominate over the surface one in transport and optical measurements, hence making very challenging to detect the topological character of those compounds.

Therefore, the target for recent research about TIs was to achieve highly resistive (low bulk carrier concentrations) crystals.

Nearly stoichiometric n-type Bi_2Se_3 crystals has a defect equilibrium that can be written as

$$Bi_2Se_3 \rightleftharpoons 2Bi_{Bi} + 3V_{Se}^{\bullet\bullet} + \tfrac{3}{2}Se_2(g) + 6e'$$

while the nearly stoichiometric Bi_2Te_3 (p type) has the following defect equilibrium

$$Bi_2Te_3 \rightleftharpoons 2Bi'_{Te} + 2h^{\bullet} + Te_2(g) + Te_{Te}$$

A transition from p type to n type behavior in the solid solution $Bi_2(Te_{1-x}Se_x)_3$ occurs (already reported 40 years ago [36]) and the bulk resistivity (see Sect. 1.3.2) can exceed 1 Ωcm.

The crystal structures of Bi_2Se_2Te and Bi_2Te_2Se are shown in Fig. 1.15. Powder X-ray diffraction (XRD) of a sample of Bi_2Te_2Se is also reported in Fig. 1.16: one can see a fully ordered structure in which the Te and Se atoms occupy their own distinct crystallographic sites with the Te in the outer chalcogen layers and the Se in the inner chalcogen layer. When the Te atoms are replaced by the more electronegative

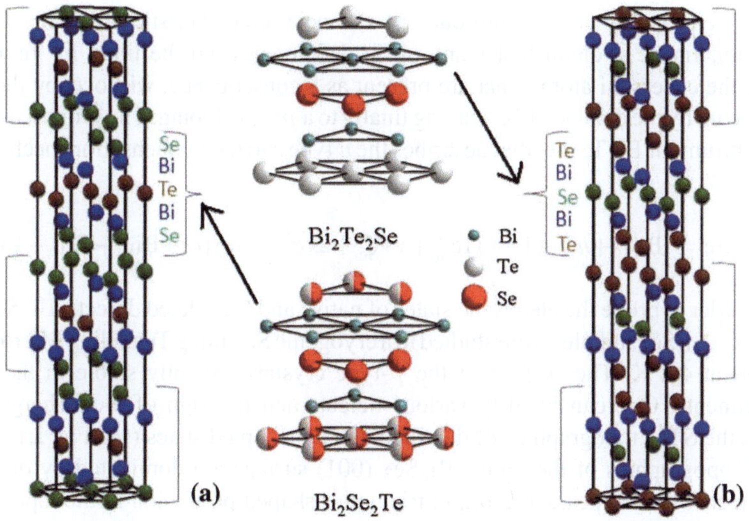

Fig. 1.15 Comparison of quintuple-layers of Bi_2Se_2Te (**a**) and Bi_2Te_2Se (**b**). Atomic positions with colored half hemispheres are occupied by mixtures of chalcogens Se and Te [18]

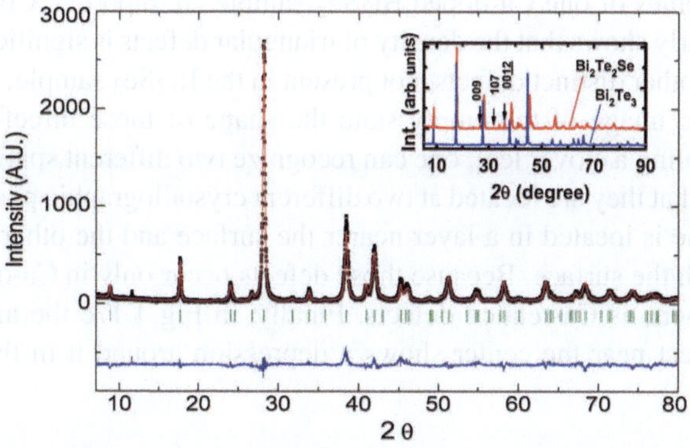

Fig. 1.16 Observed (*open circles*), calculated (*solid line*) and difference (*lower solid line*) XRD patternd of Bi_2Te_2Se [32]. *Inset* Comparison of the XRD patterns of Bi_2Te_2Se and Bi_2Te_3; *arrows* indicate the peaks characteristic of Bi_2Te_2Se [33]

Se atoms in the $Bi_2(Te_{1-x}Se_x)_3$ solid solution, the Se atoms initially fill the central layer for x=1/3 and then start to replace the Te atoms in the outside layer: this result is consistent with early XRD experiments. [32]

Actually, stoichiometric Bi_2Te_2Se is always a heavy-doped *n* type material showing metallic behavior. The slight tuning of the Te/Se ratio by changing the nominal composition at the percent level (<15 %) does not lead to semiconducting samples.

On the other hand, reducing the Se starting concentration brings to higher resistance samples: this behavior indicates that reducing Se starting concentration in

Bi_2Te_2Se probably doesn't introduce Se vacancies characteristic of Bi_2Se_3. On the contrary, Jia et. al claim that many of Se deficiencies will be filled by Te atoms, while the excess Bi atoms, that are present as a consequence, will occupy the sites left vacant by the displaced Te, leading finally to a p type doping. The proposed defect equilibrium for Bi_2Te_2Se, that describes the n type carrier compensation mechanism, is

$$\mathrm{Bi_2Te_2Se} \rightleftharpoons \mathrm{Bi'}_{\mathrm{Te}} + h^{\bullet} + (1-x)\mathrm{Te}_{\mathrm{Se}}^{\times} + x\,V_{\mathrm{Se}}^{\bullet\bullet} + 2xe' + \tfrac{1}{2}\mathrm{Se_2}(g) + \mathrm{Bi_{Bi}} + \mathrm{Te_{Te}} + \tfrac{x}{2}\mathrm{Te_2}(g)$$

In order to probe the electronic states of native and Ca-related defects, Bi_2Se_3 and $Bi_{1.98}Ca_{0.02}Se_3$ samples were studied in a cryogenic Scanning Tunneling Microscope (STM) at 4.2 K. The surface of the p-type crystals are fully stable in the STM experiments. One can identify various defects and the sign of their charge state forms the STM topographies of the filled and unoccupied states (see Fig. 1.17). The STM topographies of the native Bi_2Se_3 (001) surface are dominated by one type of defect, which appears as a bright triangular-shaped protrusion in the topography of empty states (Fig. 1.17a). Those defects are about 40 $\AA$ across on average but vary in size, indicating that they are in various layers beneath the surface. Since no other defects are observed, those defects are attributed to Se vacancies. In Fig. 1.17b, c the topographies of one Ca-doped Bi_2Se_3 sample are reported. Comparison with Fig. 1.17a clearly shows that the density of triangular defects is significantly reduced. Furthermore, other distinct defects, not present in the Bi_2Se_3 sample, are noticeable. In topographic image of the empty state the shape of those threefold symmetric defect, resembling a clover leaf, one can recognize two different spatial extents: one can conclude that they are located at two different crystallographic positions, namely the smaller one is located in a layer nearer the surface and the other one in a layer deeper beneath the surface. Because those defects occur only in Ca-doped samples, we identify them as Ca-related defects. Finally, in Fig. 1.17c the area around the triangular defect near the center shows a depression around it in the filled states,

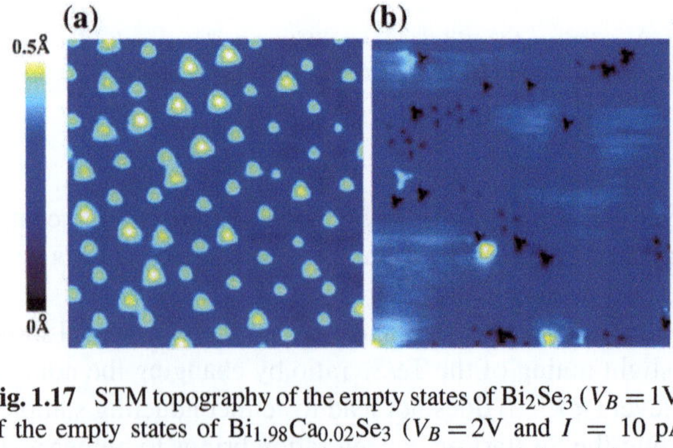

Fig. 1.17 STM topography of the empty states of Bi_2Se_3 ($V_B = 1$V and $I = 10$ pA) (**a**); topography of the empty states of $Bi_{1.98}Ca_{0.02}Se_3$ ($V_B = 2$V and I = 10 pA) (**b**); topography of the filled states over the same area in (**b**) (V_B = -1V and I = 10 pA) (**c**). All STM topographies are 500×500 $\AA^2$ [34]

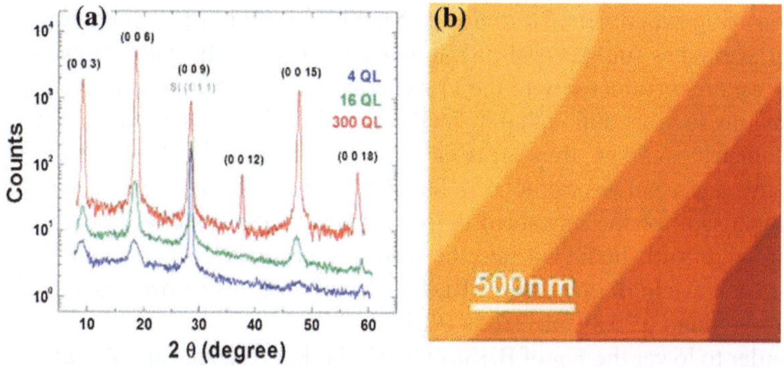

Fig. 1.18 High resolution XRD pattern of three different films of Bi_2Se_3. 1.5× 1.5 μm^2 scanned AFM image of a 300 QL thick Bi_2Se_3 film grown on Al_2O_3 (0001) (**b**) [35]

demonstrating it is positively charged. In contrast, the Ca-related defects exhibit an area of enhancement around them,indicating that they are negatively charged.

In the present work, in addition to the crystals above described, thin films of Bi_2Se_3 of different thickness have been measured.

The self-organization phenomenon renders the film thickness always an integer number of quintuple layers along the c axis. The film growers thus conclude that the film grows up by a stacking of Bi-Se quintuple layers. This is a very important self-organization behavior that makes it easy to achieve a stoichiometric film, even with not well controlled Bi and Se coverage ratio.The excess Se atoms are spontaneously desorbed during annealing, as has been proven by X-ray Photoelectron Spectroscopy (XPS) measurements before and after annealing. Furthermore, by precisely controlling the Bi/Se coverage, they have obtained atomically smooth films, with tunable thickness quintuple layer by quintuple layer. The smallest thickness of the films reaches 1 nm, namely one quintuple layer. In Fig. 1.18a the XRD results of films with different thickness of Bi_2Se_3 on a substrate of sapphire are reported: the peaks observed in the θ–2θ scan are consistent with the c-axis oriented Bi_2Se_3 phase. In Fig. 1.18b the AFM (Atomic Force Microscope) image of the same sample is shown: large terraces (largest ever reported for Bi_2Se_3 thin films) are observed, further verifying the high quality of the grown films. XRD pattern of three different films of Bi_2Se_3 is also reported in Fig. 1.18a.

1.3.2 Transport Properties

Investigating the charge transport characteristics of the surface states has proven to be a challenging, task, because the surface current is usually—by one or two orders of magnitude—smaller than the bulk current. The main difficulty about TIs, especially when using experimental techniques that do not distinguish directly between bulk and surface states (unlike ARPES), is indeed that the bulk states often contribute with a residual conductivity to the low energy and dc electrodynamics. As seen in the previous section, although pure Bi_2X_3 (X = Se, Te) is expected to be nominally stoichiometric samples, they are well known to be *n* and *p* type semiconductors owing to excess carriers introduced via Se or Te site (antisite) defects, respectively. To compensate for the unwanted defect dopants, trace amounts of carriers of the opposite sign must be added into the naturally occurring material, which may be easier to achieve in Bi_2Se_3 than in Bi_2Te_3 because the former one has a much larger bandgap (around 0.35 eV instead of 0.18 eV) [37].

In order to lower the E_F of Bi_2Se_3 into the bulk forbidden gap,one can substitute trace amounts of Ca for Bi in as-grown Bi_2Se_3 , where Ca acts as an acceptor [37] (see also Sect. 1.3.1).

In Fig. 1.19 the 2-300 K resistivities in the *ab* plane for undoped Bi_2Se_3 ($x = 0$) and $Bi_{2-x}Ca_xSe_3$ crystals for $x = 0.005$, 0.02 and 0.05 are reported, since our Ca-doped sample has a concentration of Ca corresponding to $x = 0.0002$, that is by a factor of 25 lower than the least doped of the above reported samples. Weakly metallic resistivities, quite common in high carrier concentration small band-gap semiconductors, are found. In detail, at 10 K the resistivity is in the 0.3–1.5 mΩ

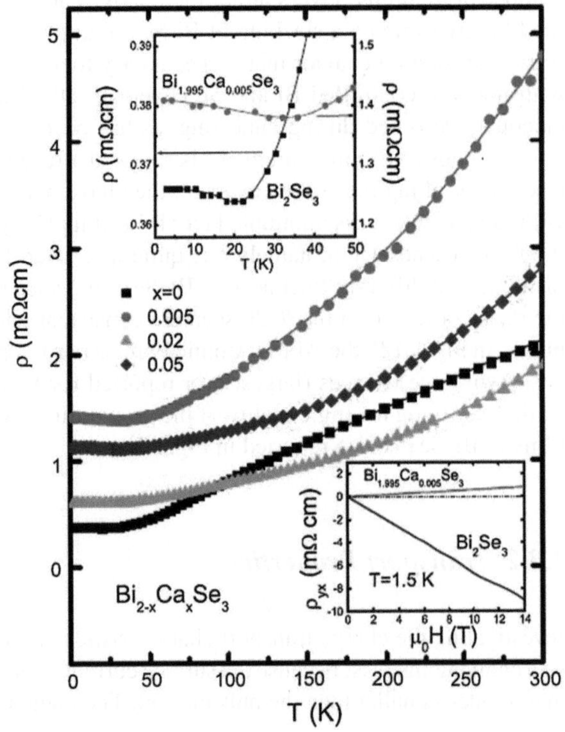

Fig. 1.19 Temperature dependent resistivity in the *ab* plane of single crystal of Bi_2Se_3 and $Bi_{2-x}Ca_xSe_3$ at specific x. The *upper inset* details the resistivity at low temperature. The *lower inset* shows its magnetic field dependence, which provides the Hall effect [34]

range. The lightly doped $Bi_{2-x}Ca_xSe_3$ has a 1×10^{19} cm^{-3} carrier concentration, determined by the Hall measurements shown in the lower inset of Fig. 1.19 [34]. As one can see in the upper panel of Fig. 1.19 the ratio $\rho_{300}/\rho_{4.2}$ is about 3 for both the undoped n type and the lightly Ca-doped crystals. The decrease in the resistivity on cooling stops below about 50 K. The Hall resistivity does not change appreciably below 20 K: this suggest that the carrier concentration is essentially constant in that temperature regime. Both crystals have been experienced also quantum oscillation (Shiubnikov de Haas effect) in the resistivity. The period of the oscillations allowed Hor et al. [34] to determine the Fermi wave vector k_F.They have estimated the metallicity parameters $k_F L$ at 4 K, combining that result with the resistivity data. It is 13 for the p type sample and 113 for the n type ones. It means that the mean free path L is 18 nm and 370 nm, respectively, well within the metallic regime.

Considering now the compensated sample Bi_2Te_2Se, one can observe its resistivity at zero magnetic field in Fig. 1.20a. Below 40 K the value of ρ has a value in the range 5–6 Ω cm, about 1000 times higher than the Bi_2Se_3. ρ can be expressed it also as an areal resistance $R_\square = \rho/d$, where d is the thickness of the sample, here equal to 110 μm, that is $R_\square = 400\Omega$ [38].

Despite the high resistance, Bi_2Te_2Se exhibits a logarithmic temperature dependence. The Hall coefficient below 10 K provides a very small bulk carrier concentration $n_b = 2.6 \times 10^{16}$ cm^{-3}, corresponding also to a low bulk mobility $\mu_b = 50$ cm^2/Vs.

Such a low mobility should not produce Shubnikov de Haas (SdH) oscillations for a magnetic field $B < 14$ T; nevertheless, as one can see in Fig. 1.20b, the Hall resistivity ρ_{xy} displays prominent SdH oscillations at 4.4 K in its derivative $d\rho_{xy}/dB$ *vs* B.

The bulk conductance and the surface conductance independently contribute to charge transport. Since the conductance is $G = 1/R_\square$ and σ_{ij} is the total conductivity, one can write

$$\sigma_{ij} = \sigma_{ij}^b + G_{ij}^s/d \tag{1.37}$$

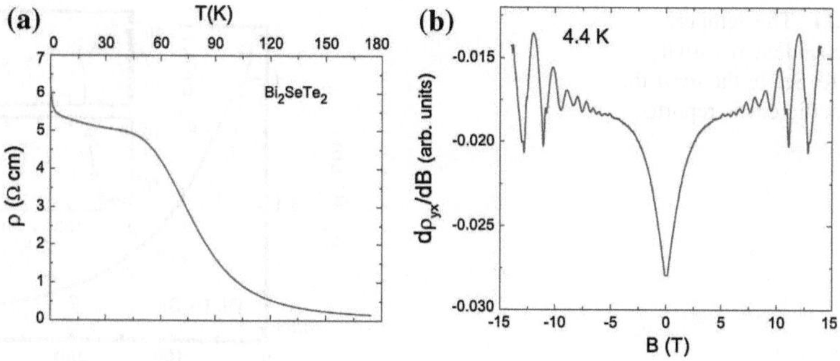

Fig. 1.20 The temperature dependent resistivity of a cleaved Bi_2Te_2Se crystal (**a**). The SdH oscillations observed in crystals of the same batch in $d\rho_{xy}/dB$ versus B at 4.4 K (**b**) [38]

where σ_{ij}^b is the bulk conductivity and G_{ij}^s is the surface conductance. If one subtracts the smooth background from the total conductivity, one obtains $\Delta\sigma_{ij}$, which, plotted *vs* $1/B$, exhibits the SdH oscillations. Oscillations decrease rapidly as T increases above 10 K. One can fit those oscillations by the formula

$$\Delta\sigma_{ij}/\sigma_{ij} = \left(\frac{\hbar\omega_c}{2E_F}\right)^{1/2} \frac{\lambda}{\sinh\lambda} e^{\lambda_D} \cos\left[\frac{2\pi E_F}{\hbar\omega_c} + \frac{\pi}{4}\right] \tag{1.38}$$

with $\lambda = 2\pi^2 k_B T/\hbar\omega_c$ and $\lambda_D = 2\pi^2 k_B T_D/\hbar\omega_c$, where ω_c is the cyclotron mass and $T_D = \hbar/2\pi k_B \tau$ is the Dingle temperature (with τ the lifetime).

The SdH frequency is hence $2\pi E_F B/\hbar\omega_c = 4\pi^2\hbar n_s/e$ where $n_s = k_F^2/4\pi$ (per spin). Equation 1.38 can be simply employed for a Dirac system if one writes the cyclotron mass as $m_c = E/v_F^2$. Xiong et al. have found that they need two SdH frequencies to fit the data, which present a beating between each other. The authors found two different surface carrier density differing by only 5 %, that is $(n_{s1}, n_{s2}) = (1.79, 1.71) \times 10^{12}\text{cm}^{-2}$, corresponding to an average Fermi wavevector $k_F = 0.047\ \AA^{-1}$ and a Fermi velocity $v_F \sim 6 \times 10^5$ m/s. From the fit parameter T_D they also found a mean free path $l = 70$–100 nm and a surface mobility $\mu_s = el/\hbar k_F = 2800 + 250 cm^2$/Vs. It is worth noting that μ_s is strongly enhanced over μ_b (by about a factor of 60). Finally, by the decrease in amplitude of the oscillations with T at fixed B, they obtained an effective mass for the carriers $m^* = 0.089 m_e$.

The high mobility sustains the SdH originated from surface states: indeed, if the period of the SdH oscillation was of the order of the Fermi wavevector, one would find a mobility compatible with a resistivity lower by about 10^4 than the measured one [38].

Independently, SdH oscillations were also observed in Bi_2Te_2Se by Ando et al. [33], who measured the resistivity reported in Fig. 1.21.

They found a surface carriers density $n_s = 1.5 \times 10^{12}$ cm^{-2}, corresponding to a Fermi wavevector $k_F = 4.4 \times 10^6$ cm^{-1} and velocity $v_F = 4.6 \times 10^7$ cm/s.

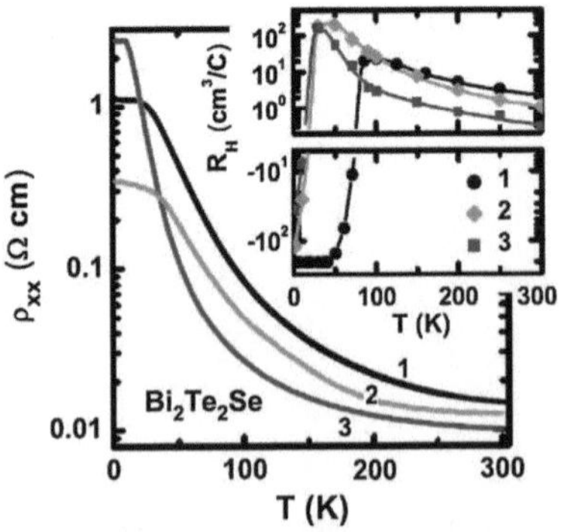

Fig. 1.21 The temperature dependent resistivity of Bi_2Te_2Se. In the *inset* the Hall coefficient is reported [33]

The mean free path is about 22 nm and the corresponding surface mobility is about 760 cm^2 Vs.

Thin Films of Bi_2Se_3

Here we report the resistance data on high quality thin film of Bi_2Se_3. Those samples have been measured in this work together with the bulk crystals described above.

The thickness of the samples, grown by Bansal et al. at Rutgers University, varies from 256 to about 8 QL (1 QL $\sim$ 1 nm).

In Fig. 1.22 the low temperature dependent sheet resistance is reported. It is worth noting that the temperature dependence is quite thickness independent for samples between 256 and $\sim$8 QL; this provides a conductance quite constant for variable thickness, suggesting that, in this thickness range, it is dominated by surface transport channels [39].

Hall measurements provide more insights regarding the origin of those surface channels. If all carriers have the same mobility, $R_{xy}(B)$ should appear as a straight line, whose slope is determined by $1/n_s e$, where n_s represents the surface carriers density. A certain nonlinearity in $R_{xy}(B)$ data suggests that some different kinds of surface carriers with comparable mobilities are present (carriers with lower order of magnitude should not affect $R_{xy}(B)$). Therefore, one can distinguish two channels of transport, with n_{ss1} and n_{s2} respectively. This model well fits to the Hall resistance data. This model considers carriers on opposite surfaces or on different bands as part of the same channel, if they have similar mobilities.

By the fitting parameters Bansal et al. have found a nearly constant sheet carrier density $n_{s1} \sim 3 \times 10^{13}$ cm^{-2}. The other channel shows a $n_{s2} \sim 8 \times 10^{12}$ cm^{-2}, at large thicknesses which gradually decrease down to $\sim$ 8 QL.

An explanation of this behavior comes from ARPES measurements. These showed that not only the topological surface states but also the 2DEG states in the quantum confined accumulation layers can contribute to surface transport. If one assumes that only the lowest level of the 2DEG is filled, considering the spin degeneracy, the sheet carrier density of the topological surface state is $n_{s,TI} = k^2_{F,TI}/4\pi$, while the one of the 2DEG is $n_{s,2DEG} = k^2_{F,2DEG}/2\pi$. Because $k_{F,2DEG} < k_{F,TI}$, one should always have $n_{s,2DEG} < n_{s,TI}$. Then, one can simply identify n_{s1} with $n_{s,TI}$ and n_{s2}

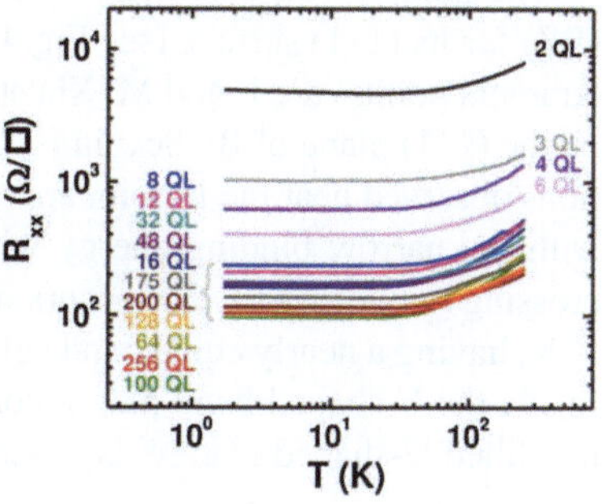

Fig. 1.22 Temperature dependent sheet resistance of different films of Bi_2Se_3 [39]

with $n_{s,2DEG}$. When the thickness of the sample approaches that of the 2DEG, the confinement starts to affect the energy levels of the 2DEG. Since the film confines with air on one side and with sapphire on the other one, the thickness confinement can be approximated by the infinite potential model. For such a model, the lowest energy level from the bottom of the conduction band is given by $h^2/8m^*t^2$, where t is the thickness of the sample and m^* its effective mass (equal to $\sim 0.11 m_e$): one finds 0.04 eV for $t = 8$ nm and 0.6 eV for $t = 2$. By comparing this with the typical band-bending energy of $0.1 \sim 0.3$ eV, reported in ARPES studies [40], the 2DEG starts to feel the thickness effect when $t \sim$ 8QL being much affected when $t \sim$ 2QL. This suggests that n_{s2}, which starts to change at ~8 QL, is more consistent with the 2DEG behavior.

Since the thickness of the topological surface state can be estimated from the ratio $\hbar v_F/E_g$, where E_g is the band gap energy (about 300 meV for the Bi_2Se_3), and is about 1 nm (1 QL), the thickness of the top and bottom surface states results ~2 nm, that is exactly the thickness of the surface channel n_{s1}.

1.3.3 Photoemission Data

In graphene, the chemistry of the carbon atoms naturally locates the Fermi level at the Dirac point (that is, the point at which the two cones intersect), where the density of states vanishes. This means that the density of carriers in graphene is highly tunable using an applied electrical field and allows for different applications of this material. The surface Fermi level of a topological insulator does not have any particular reason to be at the Dirac point; however, through a combination of surface and bulk chemical modifications, tuning to the Dirac point in Bi_2Se_3 was demonstrated [6]. This control of chemical potential is important for applications, as well as for a proposal to create a topological exciton condensate by biasing a thin film of topological insulator [41]. Known to be excellent thermoelectric materials, Bi_2Te_3 and Bi_2Se_3 have been investigated independently, particularly at Princeton, where the Hasan's group observed in ARPES experiments the single Dirac-cone surface state of Bi_2Se_3 samples, prepared by Robert Cava and coworkers [6].

Bi_2Se_3

The bulk crystal symmetry of Bi_2Se_3 is related to a hexagonal Brillouin zone (BZ) for its (111) surface (see Fig. 1.23d), whose TR invariant moment (also called Kramers points) are $\bar{\Gamma}$ and $\bar{M}$. Xia et al. [42] have performed ARPES measurements on the (111) plane of Bi_2Se_3: in Fig. 1.23a–c the electronic spectral weight distributions observed near the $\bar{\Gamma}$ point are reported. A clear V-shaped band pair is observed within a narrow binding-energy window, approaching the Fermi level (E_F). The crossing between with E_F occurs at 0.09 $\AA^{-1}$ along $\bar{\Gamma}$-$\bar{M}$ and at 0.10 $\AA^{-1}$ along $\bar{\Gamma}$-$\bar{K}$, having a nearly equal band velocities ($v_F \times 10^5$ m/s) along the two directions. Inside the V-shaped band pair, a continuum-like manifold of states, corresponding to a filled U-shaped feature, is observed.

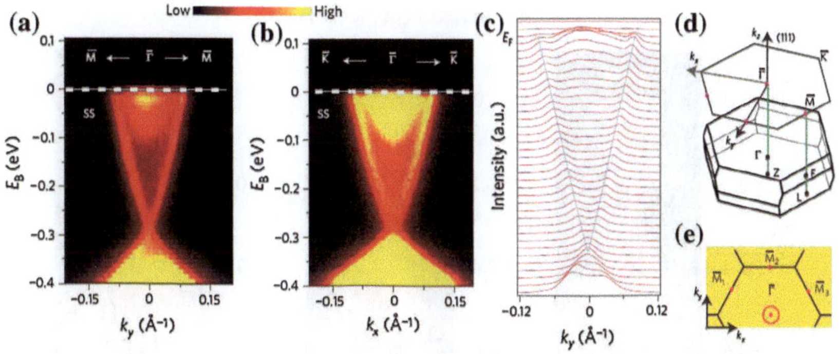

Fig. 1.23 High-resolution ARPES measurements of surface band dispersion on Bi_2Se_3 (111): electron dispersion data measured with incident photon energy of 22 eV near the $\bar{\Gamma}$-point along the $\bar{\Gamma}$-$\bar{M}$ (**a**) and $\bar{\Gamma}$-$\bar{K}$ (**b**) momentum space cuts. The momentum distribution curves corresponding to (**a**). A sketch of the full bulk 3D Brillouin zone (BZ) of Bi_2Se_3 and the 2D BZ of the projection (111) surface (**c**) [42]

Considering the bulk, one has $E(\mathbf{k},\uparrow) = E(-\mathbf{k},\downarrow)$ for TR symmetry and $E(\mathbf{k},\uparrow) = E(-\mathbf{k},\uparrow)$ for space inversion symmetry. Therefore, all the bulk bands are double degenerate. Nevertheless, since the space inversion symmetry is broken at the terminated surface, surface states (SSs) are generally spin-split on the surface by spin-orbit coupling, except at the Kramers point. Usually, in spin-orbit-coupled materials, like for instance gold, the surface states are doubly degenerate only at $\bar{\Gamma}$.

In Bi_2Se_3 the SSs emerge form the bulk continuum, cross each other at $\bar{\Gamma}$, pass through the Fermi level and eventually melt into the bulk conduction-band continuum. Then at least one continuous band-thread crosses the bulk band gap between a pair of Kramers points. Local Density Approximation (LDA) calculations showed that only if realistic Bi SOC values are included, the crossing between a singly degenerate gapless surface bands and the Fermi level is possible [42]. Furthermore, the experimental data of Fig. 1.23a–c have a rigid shift of the chemical potential with respect to the theoretical calculations: this is due to the electron doping of the Bi_2Se_3, in agreement with its n type character [42].

The experimental V-shaped pure SS band is dispersive towards E_F; moreover, inside it an electron-pocket U-shaped continuum is observed near E_F. Such filled U-shaped broad feature roughly corresponds to the bottom of the conduction band, in agreement with the n-type character of the material and to the theoretical calculations.

In Fig. 1.24a a more detailed photon-energy dependence of all the band features is reported. A modulation of incident photon energy allows one to probe the k_z dependence of some bands and hence to distinguish surface from bulk contributions to a given ARPES signal. The data don't show a strong k_z dispersion of the lowest energy bands on the "U" pocket, although the full continuum has some dispersion. The electron lack of strong dispersion suggests that the inner pocket continuum features are probably a mixture of surface-projected conduction band states, which also includes some band-bending effects near the surface and the full continuum of

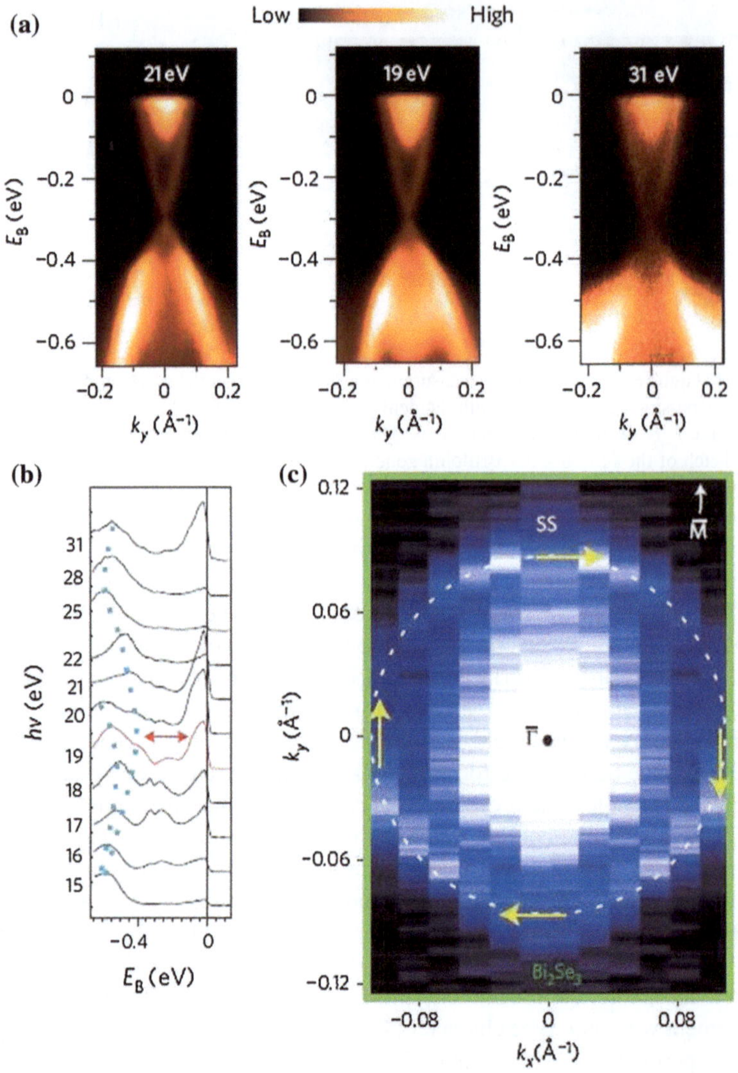

Fig. 1.24 The energy dispersion data of Bi_2Se_3 along the $\bar{\Gamma}$-$\bar{M}$ cut measured with the photo energy of 21, 19 and 31 eV (**a**). The energy distribution curves obtained from the normal-emission spectra measured using 15–31 eV photon energies reveal two dispersive bulk bands below −0:3 eV (*blue dotted lines*)(**b**). High momentum resolution data around $\bar{\Gamma}$ reveal a single ring formed by the pure SS V-shaped Dirac band (**c**) [42]

bulk conduction band states of layers just beneath the surface [42]. Similar behavior is also observed in other semiconductors [43].

In a study of the k_z dependence (see Fig. 1.24b), two bands having energies below −0.3 eV (blue dotted lines) are observed: they reflect the bulk valence bands, in addition to two other non-dispersive features associated with the two sides of the pure SS Dirac bands. The red curve instead has been measured exactly at Γ point and suggests that the Dirac point lies inside the bulk bandgap: considering the bottom of the "U" band as the minimum of the bulk conduction band, Xia et al. have estimated that a bandgap of about 0.3 eV occurs in the bulk of the undoped material [42].[3] Then it is supposed that the magnitude of band bending near the surface is not larger than 0.05 eV. It is worth noting that in Bi_2Se_3 the Fermi level should lie deep inside the gap and only pure surface bands contribute to surface conduction: the topological character of a purely insulating Bi_2Se_3 would be determined no matter the "U" feature.

In Fig. 1.25a the complete surface FS (Fermi surface) map is shown. The observed features in the 2D (111) surface BZ are all centered around $\bar{\Gamma}$. None of the TR invariants points, located at $\bar{M}$, are enclosed by any FS. In Fig. 1.24c the detailed spectral behavior, obtained with high momentum resolution, around $\bar{\Gamma}$ is reported: one can see a singly degenerate ring-like feature formed by the outer "V" pure SS

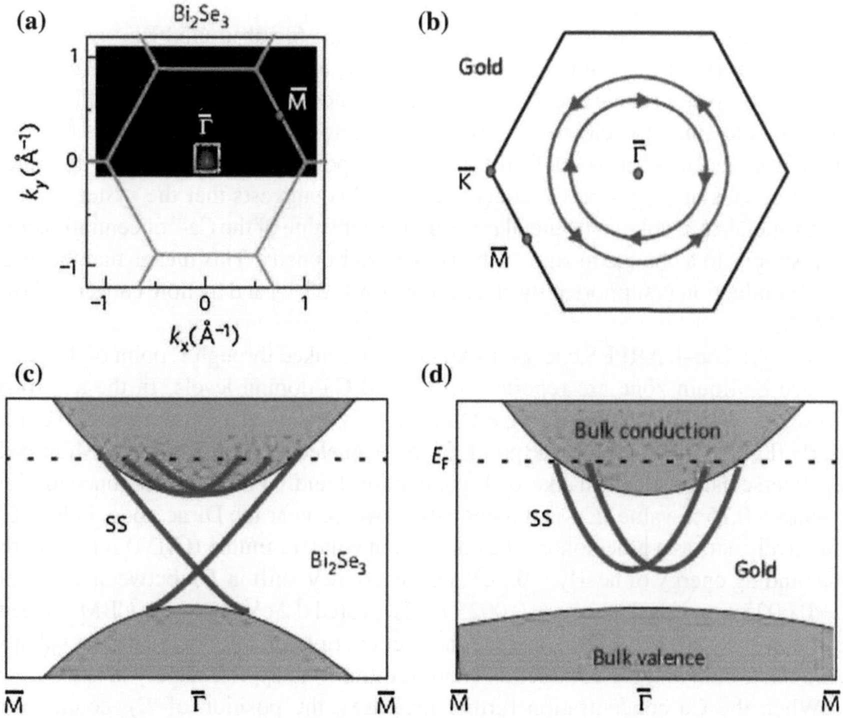

Fig. 1.25 Comparison between Bi_2Se_3 and gold. The observed surface FS of Bi_2Se_3 (**a**). The surface FS of Au (111) (**b**). Schematic SS topologies in Bi_2Se_3 and gold (**c**) and (**d**) [42]

[3] Such a value of the bulk gap is in agreement with the one estimated from bulk measurements or calculations.

band, which corresponds to a horizontal cross section of the upper Dirac cone in Fig. 1.23a, surrounds the conduction band continuum centered at $\bar{\Gamma}$. As seen in Sect. 1.2.5, if an electron encircles the surface FS that enclose a Kramers point, it obtains a geometrical quantum phase (Berry phase) of $\pi \mathrm{mod} 2\pi$ in its wavefunction.

In the insulating Bi_2Se_3 the chemical potential (Fermi level) should lie inside the bandgap, then its surface should carry a global $\pi \mathrm{mod} 2\pi$ Berry phase. In most SOC materials, like gold (Au (111)), the surface FS is formed by two spin-orbit-split rings, generated by two singly degenerate parabolic bands (not Dirac like), that are shift from each other in momentum space, both of which enclose the $\bar{\Gamma}$ point (see Fig. 1.25d, b). Such a FS topology leads to a 2π or 0 Berry phase, since the phases from the two rings add or cancel each other: gold-like topology is trivial (despite its SOC origin), while that of Bi_2Se_3 is not [42].

According to those ARPES data about Bi_2Se_3, it is possible to obtain a fully undoped compound by its chemical hole-doping in order to shift the chemical potential inside the bulk gap: this is what has been done for the chemically doped and compensated compounds described in the next section.

$Bi_{2-x}Ca_xSe_3$, Bi_2Se_2Te and Bi_2Te_2Se

In order to lower the E_F of Bi_2Se_3, Hsieh et al. [6] substituted Bi^{3+} with Ca^{2+} in as-grown Bi_2Se_3, since Ca acts as an acceptor, as shown in Scanning Tunneling Microscopy (see previous section). They performed time-dependent ARPES measurements to study the electronic structure evolution of $Bi_{2-x}Ca_xSe_3$ as a function of Ca doping, in order to confirm the data transport, that show a resistivity which sharply peaks at 0.25 % in Ca concentration. This suggests that the system undergoes a metal-to-insulator-to-metal transition. That value of the Ca-concentration also corresponds to a change in sign of the Hall carrier density. This means that the electrical conduction is supported by electron (below 0.25 %) and by hole carriers (above 0.25 %).

In Fig. 1.26c–h ARPES energy dispersion maps taken through $\bar{\Gamma}$ point of the (111) surface Brillouin zone are reported for several Ca-doping levels. In the as grown Bi_2Se_3 (x = 0, δ in figure) a single Dirac cone is observed and the Fermi level lies nearly 0.3 eV above at the Dirac point, forming an electron Fermi surface. Moreover, E_F intersects the electron-like bulk conduction band. When the Ca-concentration reaches a 0.25 % value, E_F is dramatically lowered near the Dirac node, indeed Ca effectively acts as a hole donor. The conduction band minimum (CBM) for x = 0 lies at a binding energy of nearly −0.1 eV, hence a 0.3 eV shift in E_F between x = 0 and x = 0.0025 suggests that for x = 0.0025 E_F is located 0.2 eV below the CBM. Indeed, E_F is in the bulk bandgap, since the indirect energy gap (gap between the valence band maximum and the conduction band minimum) is approximately 0.35 eV.

When the Ca concentration further increases, the position of E_F continues a downward trend, that for x = 0.01 it is located clearly below the Dirac cone and intersects the hole-like bulk valence band. The systematic lowering of the Fermi level with increasing Ca concentration in $Bi_{2-x}Ca_xSe_3$ agrees with the measured transport behaviour.

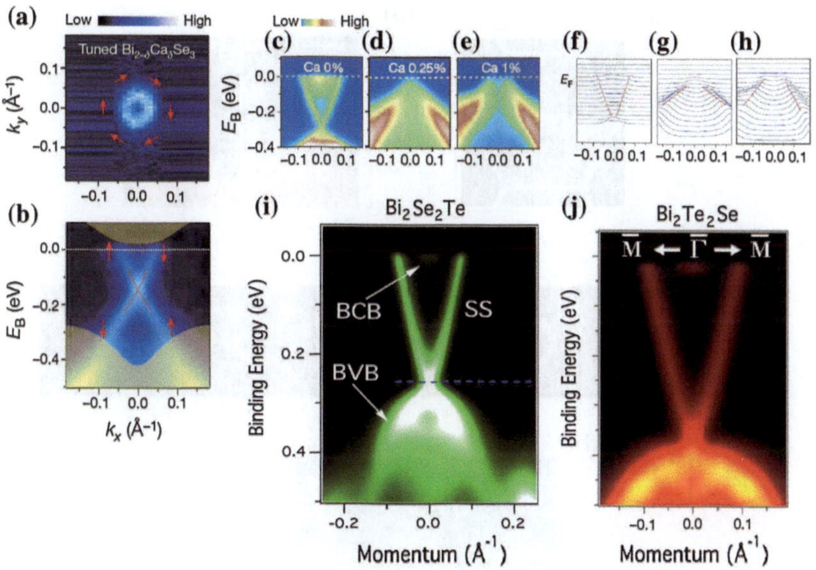

Fig. 1.26 ARPES intensity map at E_F of the (111) surface of tuned stoichiometric $Bi_{2-x}Ca_xSe_3$ (**a**)and the related ARPES dispersion (**b**). ARPES band dispersion images of $Bi_{2-x}Ca_xSe_3(111)$ through $\bar{\Gamma}$ collected within 20 min after cleavage for x (δ in figure) = 0, 0.0025 and 0.01 (**c**), (**d**) and (**e**), respectively. Corresponding momentum distribution curves (*red line* are guides to the eye) (**f**)–(**h**) [37]. ARPES energy maps along the $\bar{\Gamma}$-$\bar{M}$ momentum direction for Bi_2Se_2Te (**i**) [44]. ARPES energy dispersion of Bi_2Te_2Se along the $\bar{\Gamma}$-$\bar{M}$ direction (**l**) [18]

Considering now one of the two compensated sample studied in this work, that is Bi_2Se_2Te, one can see in Fig. 1.26i its ARPES energy dispersion: it provides a single Dirac cone with a bulk insulating gap of about 250 meV. The Fermi level of this system lies in the bulk conduction band (BCB) and the valence band (BVB) is located below the Dirac point.

The other analyzed compensated sample is Bi_2Te_2Se. In Fig. 1.26l its ARPES characterization of the band structure along the $\bar{\Gamma}$-$\bar{M}$ is shown. The single Dirac point is buried below the top of the valence by approximately 50 meV, being the bulk bandgap of about 300 meV at Γ point [18]. Band structure measurements are presented by scanning over the full exagonal BZ (see Fig. 1.27a) and high resolution dispersion maps along the $\bar{\Gamma}$-$\bar{M}$ and $\bar{\Gamma}$-$\bar{K}$ directions, which trace a clear single Dirac cone, are shown in Fig. 1.27b. No other band features are observed inside the Dirac cone, suggesting that the bulk conduction band minimum is above the chemical potential and the Fermi level lies inside the band-gap.

Incident photon energy dependence on Bi_2Te_2Se, shown in Fig. 1.27d clearly confirms the surface origin of the "V" shaped Dirac band, since it doesn't show any k_z dispersion with incident photon energy. On the contrary, band structure below the Dirac point dramatically changes with k_z: this indicates that it represents the bulk bands. The Fermi momentum and velocity in that compensated compound, observed

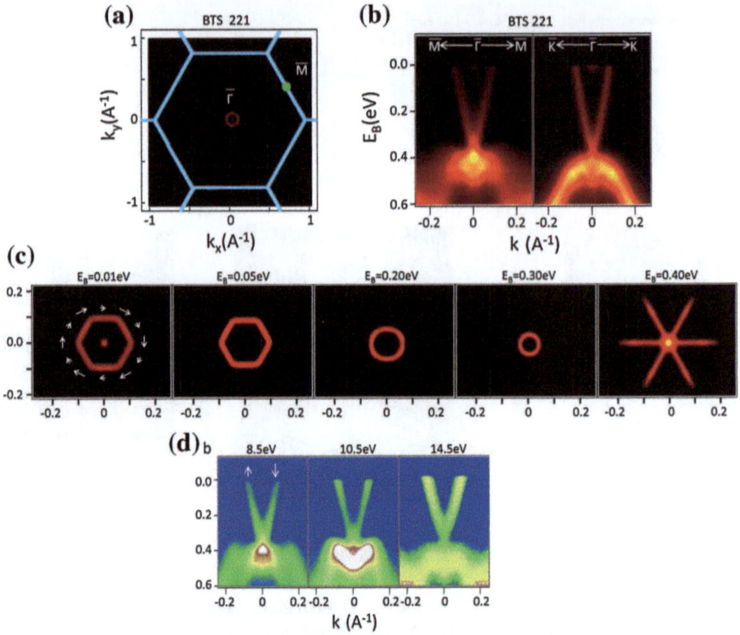

Fig. 1.27 ARPES measurements over the first BZ on Bi_2Te_2Se (**a**) and the related high resolution band dispersion (**b**). ARPES measurements of constant energy at different binding energies on Bi_2Te_2Se. In-plane spin texture from the theoretical calculation is drawn on the first panel (**c**). Incident photon energy dependence of measurements along the direction $\bar{\Gamma}$-$\bar{M}$ on Bi_2Te_2Se (**d**) [45]

by Xu et al. [45], are 0.09 $\mathring{A}^{-1}$ and 1.1×10^6 m/s, respectively, along $\bar{\Gamma}$-$\bar{M}$ and 0.08 $\mathring{A}^{-1}$ and 1.5×10^6 m/s along $\bar{\Gamma}$-$\bar{K}$. The latter one is nearly a factor of three larger than the Fermi velocity measured in any other known topological insulator.

1.3.4 Optical Properties: Previous Results of Bi_2Se_3

IR spectroscopy is a powerful tool to probe the dynamics of the charge carriers, since it allows one to study the electronic low-energy excitations and their interaction with other excitations of the systems (such as phonons). Therefore in principle it allows one to reveal signatures of the topological states, first because they are sensitive to the application of an electromagnetic field; secondly, because a magnetic-IR study can probe the energy dispersion as a function of the cyclotron resonance, that defined the Landau levels of the 2DEG in TIs.

Basov and coworkers first measured by FTIR (Fourier Transform Infrared) Spectroscopy the reflectivity and transmittance of the topological insulator Bi_2Se_3, as a function of temperature (from 6 to 295 K) and magnetic field (from 0 to 8 T). They found a narrowing of the Drude conductivity due to reduced quasi-particle scattering and an increase in the absorption edge due to direct electronic transition with lowering of temperature. Moreover one of two far-IR optical phonons is strongly renormalized and asymmetrically broaden with increasing of the magnetic field parallel to the c-axis of the layered material, suggesting the presence of an interaction between the phonon and a continuum free-carrier spectrum and of a significant magnetoelectric coupling. In the magnetic field perpendicular to the c -axis configuration, Basov et al. observed an enhancement of electronic absorption and a slight shift of the plasma edge to higher energies [30].

In Fig. 1.28b the transmittance data at zero magnetic field are reported. The thin cleaved crystal is transparent only within a narrow frequency range in the mid-IR, extending from 470 to 2000 cm^{-1}. The low-frequency absorption shows a free carrier transport. The high-frequency part refers to the direct allowed transitions from the valence band to the Fermi level. In particular, the low frequency transmission onset becomes very steep at low temperatures, together with the displacement of the mid-IR absorption edge to higher frequencies, partly due to a reduction in thermal broadening.

As the authors showed, this interpretation of the transmission data is confirmed by the reflectivity ones, reported in Fig. 1.28a. The plasma edge moves to lower frequencies and sharpens as the temperature decreases to 6 K. Two phonon absorption lines are observed: one near 61 cm^{-1} (α mode) and one near 133 cm^{-1} (β mode). They correspond to the relative transverse motion of charged Bi and Se ions (modes of E_u symmetry) and both sharpen at low temperature. A peak appears in the mid-IR

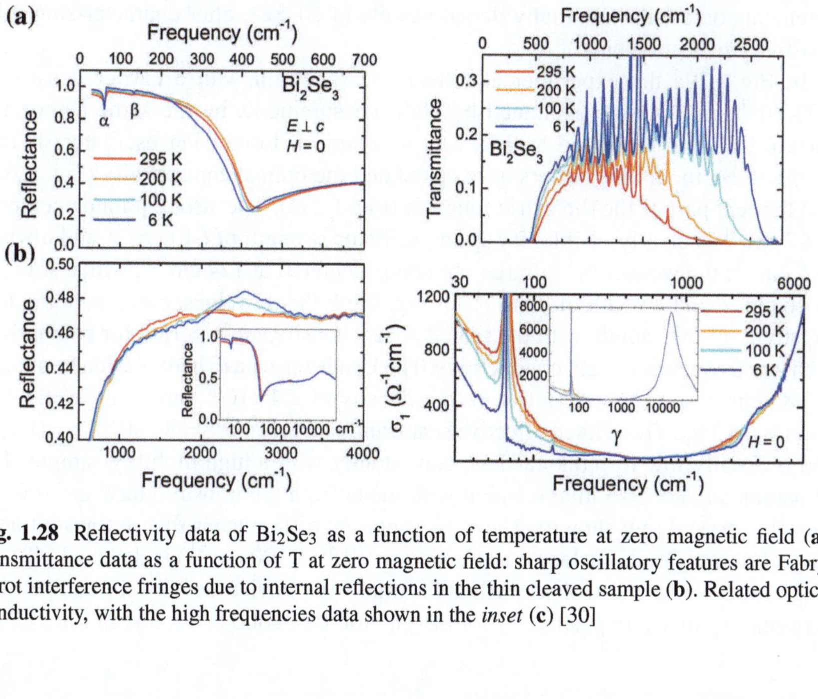

Fig. 1.28 Reflectivity data of Bi_2Se_3 as a function of temperature at zero magnetic field (**a**); transmittance data as a function of T at zero magnetic field: sharp oscillatory features are Fabry-Perot interference fringes due to internal reflections in the thin cleaved sample (**b**). Related optical conductivity, with the high frequencies data shown in the *inset* (**c**) [30]

near 2400 cm^{-1}: it is known as the Burnstein shift, that is associated with the direct allowed interband transitions [46]. Reflectivity data for the extended frequency range are reported in the inset of Fig. 1.28a.

Figure 1.28c shows the optical conductivity, obtained by Kramers-Kronig transformation of the reflectivity. A Drude term is clearly visible: it strongly sharpens with decreasing temperature. Such a sharpening reveals a broad electronic mode, labeled as γ in figure, centered in the far-IR near 300 cm^{-1}. At higher frequencies, the Burnstein shift leads to a significant transfer of spectral weight from a gapped region spanning from 1000 $cm^{-1} \div 2500$ cm^{-1} to 3000 cm^{-1}. The largest feature in the conductivity is the triplet of interband absorptions centered at 16000 cm^{-1} [47].

The phonons also sharpen and exhibit frequency shifts. In particular, the α mode has an asymmetric Fano line shape (see for details Sect. 2.6.2). This is a signature of the interaction between a discrete mode (phonon in this case) and a continuum of excitations, that is of the electron-phonon coupling. The authors have fitted a function of the Fano parameter, as described in Sect. 2.6.2, to the optical conductivity. The evolution of the fit parameters for the α and Drude modes with magnetic field parallel to the c-axis yields insight into the nature of free electron-lattice coupling in Bi_2Se_3. First of all, the linewidth of the Drude peak (γ^{Drude}) increases with field; the width of the phonon, modified by the interaction, increases linearly with field over the entire range. The Fano parameter, q, scales with $1/\tau^{Drude}$: this suggests that the strength of coupling between the phonon and the Drude term may increase as the free-carrier scattering frequency approaches the one of the lattice vibration [30]. Moreover, the negative value of q indicates that the phonon interacts with electronic absorption centered at higher frequency: an interaction with the γ mode is hence possible.

Butch et al. [48] have performed infrared reflectivity and transmittance measurements on a bulk nominally doped sample of Bi_2Se_3, after characterizing it by resistivity measurements.

In Fig. 1.29a the experimental reflectivity spectrum with a carrier density of 3.7×10^{17} cm^{-3} (value estimated by Hall measurements by the same authors) is shown. Data are presented at 6 K; a fit to a sum of Lorentzian oscillators is also reported. By the fit the authors have calculated the optical conductivity (Fig. 1.29b) and the real part of the dielectric function (Fig. 1.29c). The strong phonon centered at 67 cm^{-1} is clearly visible: it has an oscillator strength of 640 cm^{-1} and a width of 5 cm^{-1}; the weaker far-infrared phonon is centered at 124 cm^{-1}, with a strength of 76 cm^{-1} and a width of 2 cm^{-1}. In Fig. 1.29b the dc values, corresponding to a resistivity of the sample with the same carriers density, are reported for comparison with low frequency optical conductivity. The Drude term has a bare plasma frequency of 382 cm^{-1}, corresponding to a carrier density of 2.4×10^{17} cm^{-3}, if the effective mass is $0.15m_e$. The related effective scattering rate is $\gamma^{Drude} = 8$ cm^{-1} ($\tau = 0.8$ ps) and is usually low for bulk carriers, consistently with a high mobility sample. The fit parameters are also in agreement with those from the transmittance spectra of a very thin crystal (not shown). The only important difference is that in transmittance data a broader Drude surface contribution ($\gamma^{Drude} = 100$ cm^{-1}) is detected [48].

The authors have also observed that for small bulk carriers density, the competition between negative and positive ϵ_1 for the phonon (see solid black line in Fig. 1.29c)

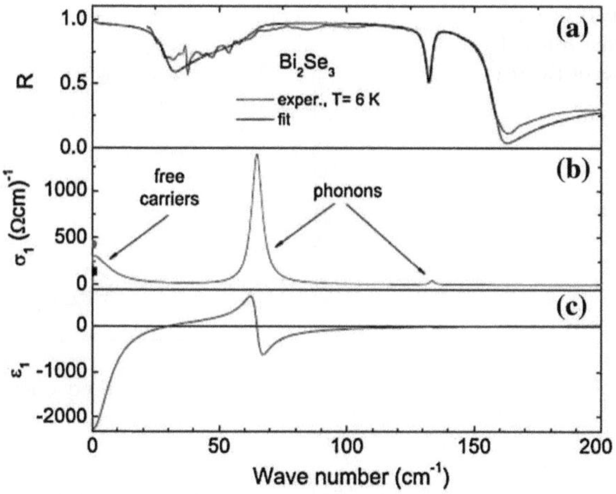

Fig. 1.29 Spectra of reflectivity (**a**), optical conductivity (**b**) and real part of the dielectric function of Bi_2Se_3 (**c**). Curves (**b**) and (**c**) are calculated from the fit to the reflectivity data in (**a**). *Blue circles* and *red squares* in (**b**) are dc values for a sample with carriers density of 2.4×10^{17} cm^{-3} [48] (Color figure online)

gives rise to an additional low frequency plasma edge ($\epsilon_1 = 0$), that leads to a transmission window near 30 cm^{-1}. The strong bismuth-dominated low frequency phonon, indeed, provides a large static dielectric function $\epsilon_1(0) = 100$. This implies a large reduction in the Coulomb potential, that is in the scattering rate from ionized impurities: this can explain the small scattering rate of the bulk carriers and hence the good metallic behavior of Bi_2Se_3. A recent publication of Armitage's group reports THz response of thin films of Bi_2Se_3 on sapphire (Al_2O_3) substrate. Those same samples are the ones used for the plasmonic investigation by us and measured by FTIR Spectroscopy in this thesis.

As already reported in Sect. 1.3.2, those films exhibit a 2D behavior for thickness less than 300 QL , because of their low bulk carrier density and high mobility.

The authors have performed Time Domain Terahertz Spectroscopy (TDTS) and in Fig. 1.30 the real part of the longitudinal conductance ($G_{xx} = \sigma_{xx}t$, where t is the thickness and σ_{xx} the longitudinal conductivity) for different thickness is reported at 6 K and zero magnetic field. The data show a clear free electron behavior, that is a narrow Drude peak (about 1.2 THz), and an optical phonon close to 2 THz. As one can see, the data are very similar to those ones of crystals described above and in Ref. [30]. The data were fitted to two identical Drude terms, one for each surface, and to a Drude-Lorentz term for the bulk phonon: these fits allowed the authors to obtain a parametrization of the data, showing that the low-frequency transport is dominated by a single charge channel. The most important information that one can deduce by those data is that the surface Drude term has an almost thickness-independent character, while the intensity of the optical phonon shows a linear increase with the thickness, typical of a bulk response. Therefore, one can conclude that the surface

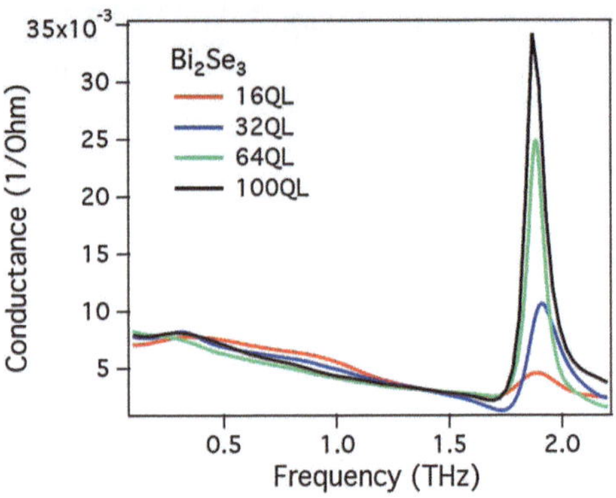

Fig. 1.30 Real part of the conductance of four films of different thickness of Bi_2Se_3 [49]

transport dominates the free electron response of those films [49] This issue is also confirmed by transport measurements on the same samples [39].

The observed scattering rate of the carriers, together with the Fermi velocity obtained by ARPES measurements of Ref. [6], gives a long mean free path of about 0.5 μm.

Furthermore, the authors have performed measurements of Kerr and Faraday rotation angles, using the time structure of TDST[4] and the sapphire substrate itself as an optical resonator with a magnetic field (see for details Ref. [50]), in order to extract the effective mass of the 2D Dirac electrons. Indeed, from the expression of the Kerr angle

$$tan(\phi_K) = \frac{2nZ_0G_{xy}}{n^2 - 1 - 2Z_0G_{xx} - Z_0^2(G_{xx}^2 + G_{xy}^2)} \tag{1.39}$$

where n is the refractive index of the substrate, Z_0 is the vacuum impedance (377 Ohm) and G_{xy} is the Hall conductance, using parameters of the fit to the total conductance, they have found an effective cyclotron mass $m^* \sim 0.35m_e$. This is a bit larger than that ($\sim 0.15m_e$) conventionally used. An estimation of the Dirac fermion effective mass calculated by Fermi energy of $\sim$ 0.5 eV (from a carrier density $n \sim 3.3 \times 10^{13}$ cm^{-2}) and Fermi velocity from ARPES data (Ref. [6]) gives only a 10 % difference from that value.

The authors conclude that the THz response is strongly independent of the bulk contribution to the conductance and the observed colossal Kerr rotation is intrinsic to the surface metallic states. Its large value is due to a near cancellation of the conductance terms in the denominator of Eq. 1.39 because of the large G_{xy} and also to an enhancement due to the reflection occurring inside the sapphire, that is at the interface of the topological insulator.

[4] The fact that TDTS measurements are resolved in time allows the separation of the different contributions to the rotation angle; this type of separation is generally not possible with standard continuous wave techniques.

References

1. C.L. Kane, E.J. Mele Phys, Rev. Lett. **95**, 146802 (2005)
2. C.L. Kane, E.J. Mele Phys, Rev. Lett. **95**, 226801 (2005)
3. B.A. Bernevig, S.C. Zhang, Phys. Rev. Lett. **96**, 106802 (2006)
4. M. König, S. Wiedmann, C. Bernevig, A. Roth, H. Buhmann, L. W. Molenkamp, X.L. Qi, S.C. Zhang, Science, 5851 (2007)
5. L. Fu, C.L. Kane, Phys. Rev. B **76**, 045302 (2007)
6. D. Hsieh, D. Qian, L. Wray, Y. Xia, Y.S. Hor, R.J. Cava, M.Z. Hasan, Nature **452**, 970 (2008)
7. H. Zhang, C.X. Liu, X.L. Qi, X. Dai, Z. Fang, S.C. Zhang, Nat. Phys. **5**, 438 (2009)
8. D.C. Tsui, Rev. Mod. Phys. **71**, 891 (1999)
9. M.Z. Hasan, C.L. Kane, Rev. Mod. Phys. **82**, 3045 (2010)
10. J.E. Moore, Nature **464**, 194 (2010)
11. F.D.M. Haldane, Phys. Rev. Lett. **61**, 2015 (1988)
12. D.P. DiVincenzo, E.J. Mele, Phys. Rev. B **29**, 1685 (1985)
13. R. Jackiw, C. Rebbi, Phys. Rev. D **13**, 3398 (1976)
14. L. Fu, C.L. Kane, Phys. Rev. B **74**, 195312 (2006)
15. C.L. Kane, J.E. Moore, Physics, World 32 (2011)
16. D. Spirito, Ph.D. Thesis, 2011
17. K. von Klitzing, G. Dorda, M. Pepper, Phys. Rev. Lett. **45**, 494 (1980)
18. H. Ji, J.M. Allred, M.K. Fuccillo, M.E. Charles, M. Neupane, L.A. Wray, M.Z. Hasan, R.J. Cava, Phys. Rev. B **85**, 201103(R) (2012)
19. B. Das, D.C. Miller, S. Datta, R. Reifenberger, W.P. Hong, P.K. Bhattacharya, J. Singh, M. Jaffe, Phys. Rev. B **39**(2), 1411 (1989)
20. M. Nakahara, *Geometry, Topology and Physics* (Adam Hilger, Bristol, 1990)
21. M.V. Berry Proc, R. Soc. Lond. A **392**, 45 (1984)
22. S. Murakami, N. Nagaosa, S.C. Zhang, Phys. Rev. Lett. **93**, 156804 (2004)
23. X.L. Qi, S.C. Zhang, Phys. Today **63**(1), 33 (2010)
24. C. Wu, B.A. Bernevig, S.C. Zhang, Phys. Rev. Lett. **96**, 106401 (2006)
25. C. Xu, J.E. Moore, Phys. Rev. B **73**, 045322 (2006)
26. A.P. Schnyder, S. Ryu, A. Furusaki, A.W.W. Ludwig, Phys. Rev. B **78**, 195125 (2008)
27. D.M. Rowe, *Handbook of Thermoelectrics*, 6th edn. (CRC Press, Boca Raton, 1995)
28. L.L. Wang, D.D. Johnson, Phys. Rev. B **83**, 241309(R) (2011)
29. E.M. Black, L. Conwell, L. Seigle, C.W. Spencer, J. Phys. Chem. Solids **2**, 240 (1957)
30. A.D. LaForge, A. Frenzel, B.C. Pursley, T. Lin, X. Liu, J. Shi, D.N. Basov, Phys. Rev. B **81**, 125120 (2010)
31. S.M. Young, S. Chowdhury, E.J. Walter, E.J. Mele, C.L. Kane, A.M. Rappe, Phys. Rev. B **84**, 085106 (2011)
32. S. Jia, H. Ji, E. Climent-Pascual, M.K. Fuccillo, M.E. Charles, J. Xiong, N.P. Ong, R.J. Cava, Phys. Rev. B **84**, 235206 (2011)
33. Z. Ren, A.A. Taskin, S. Sasaki, K. Segawa, Y. Ando, Phys. Rev. B 82, 241306(R) (2010)
34. Y.S. Hor, A. Richardella, P. Roushan, Y. Xia, J.G. Checkelsky, A. Yazdani, M.Z. Hasan, N.P. Ong, R.J. Cava, Phys. Rev. B **79**, 195208 (2009)
35. Y.S. Kim, M. Brahlek, N. Bansal, E. Edrey, G.A. Kapilevich, K. Iida, M. Tanimura, Y. Horibe, S.W. Cheong, S. Oh, Phys. Rev. B **84**, 073109 (2011)
36. S. Nakajima, J. Phys. Chem. Solids **24**, 479 (1963)
37. D. Hsieh et al., Nature **460**, 1101 (2009)
38. J. Xiong, A.C. Petersen, D. Qu, R.J. Cava, N.P. Ong, Physica E **44**, 917 (2012)
39. N. Bansal, Y.S. Kim, M. Brahlek, E. Edrey, S. Oh, Phys. Rev. Lett. **109**, 116804 (2012)
40. M. Bianchi, D. Guan, S. Bao, J. Mi, B.B. Iversen, P.D.C. King, P. Hofmann, Nat. Commun. **1**, 128 (2010)
41. B. Seradjeh, J.E. Moore, M. Franz, Phys. Rev. Lett. **103**, 066402 (2009)
42. Y. Xia, D. Qian, D. Hsieh, L. Wray, A. Pal, H. Lin, A. Bansil, D. Grauer, Y.S. Hor, R.J. Cava, M.Z. Hasan, Nature **5**, 398 (2009)

43. M. Hoesch, M. Muntwiler, V.N. Petrov, M. Hengsberger, L. Patthey, M. Shi, M. Falub, T. Greber, J. Osterwalder, Phys. Rev. B **69**, 241401 (2004)
44. M. Neupane, S.Y. Xu, L.A. Wray, A. Petersen, R. Shankar, N. Alidoust, C. Liu, A. Fedorov, H. Ji, J.M. Allred, Y.S. Hor, T.R. Chang, H.T. Jeng, H. Lin, A. Bansil, R.J. Cava, M.Z. Hasan, Phys. Rev. B **85**, 23406 (2012)
45. S.Y. Xu, L.A. Wray, Y. Xia, R. Shankar, A. Petersen, A. Fedorov, H. Lin, A. Bansil, Y.S. Hor, D. Grauer, R.J. Cava, M.Z. Hasan, arxiv:1007.5111 (2010)
46. H. Köhler, J. Hartmann, Phys. Status Solidi **63**, 171 (1974)
47. D.L. Greenaway, G. Harbeke, J. Phys. Chem. Solids **26**, 1585 (1965)
48. N.P. Butch, K. Kirshenbaum, P. Syers, A.B. Sushkov, G.S. Jenkins, H.D. Drew, J. Paglione, Phys. Rev. B **81**, 241301(R) (2010)
49. R. Valdés Aguilar, A.V. Stier, W. Liu, L.S. Bilbro, D.K. George, N. Bansal, L. Wu, J. Cerne, A.G. Markelz, S. Oh, N.P. Armitage, Phys. Rev. Lett. **108**, 087403 (2012)
50. R. Valdés Aguilar, A.V. Stier, W. Liu, L.S. Bilbro, D.K. George, N. Bansal, L. Wu, J. Cerne, A.G. Markelz, S. Oh, N.P. Armitage, Supplemental materials for. Phys. Rev. Lett. **108**, 087403 (2012)

Chapter 2
Experimental Technique, Sample Fabrication and Models for Data Analysis

Abstract In this chapter the experimental technique used is described. The Fourier Transform Infrared Spectroscopy allows one to perform reflectivity and transmittance measurements as a function of frequency in a wide spectral range, from the far infrared to the ultraviolet. Through Kramers-Kronig transformations it is possible to calculate the linear optical response functions of the material. In this sections the experimental apparatus is also described and the models used to analyze both the reflectivity and transmittance data are discussed. The last part is dedicated to a brief description of the fabrication of plasmonic devises based on TI thin films and to the fitting of the data.

2.1 Fourier Transform Infrared Spectroscopy

Fourier Transform Infrared Spectroscopy (FT-IR) is a powerful technique to obtain a rapid and broad-range spectral analysis of the electromagnetic radiation, from the far infrared ($\sim$ 1meV) to the ultraviolet ($\sim$ 5eV) with high spectral resolution. The spectrometer used is a Michelson Interferometer, whose scheme is shown in Fig. 2.1. From a point source, using collimating mirrors, the electromagnetic radiation is sent to a beam-splitter (BS) that separates it into two components. Those components are sent one to a steady mirror and the other one to a scanning mirror. Thereafter the signals are recombined in the BS and focused on the detector. The intensity, which is detected as a function of the optical path difference δ generated by the scanning of the moving mirror of the interferometer, is the interferogram $I_R(\delta)$ and contains implicitly the whole frequency dependence of the light. The Fourier transform of $I_R(\delta)$ permits to reconstruct the frequency spectrum of the light $S(\omega)$ through the relation

$$B(\omega) \propto \int \left[I_R(\delta) - \frac{1}{2} I_R(0) \right] \cos(2\pi\omega\delta) d\delta \tag{2.1}$$

P. Di Pietro, *Optical Properties of Bismuth-Based Topological Insulators*,
Springer Theses, DOI: 10.1007/978-3-319-01991-8_2,

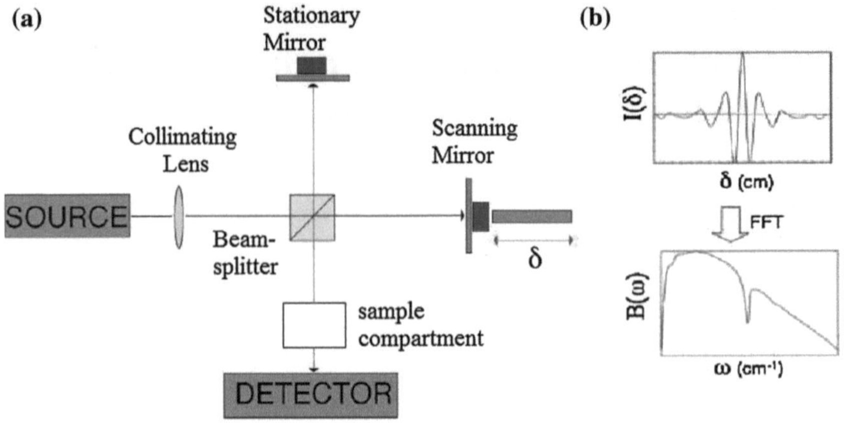

Fig. 2.1 Schematics of a Michelson Interferometer (**a**). The interferogram and the corresponding power spectrum obtained through a Fast Fourier Transform (FFT) procedure are shown (**b**)

where ω is in cm^{-1} (wavenumbers). In the case of a monochromatic light source, the interferogram is a periodic sine function and the spectrum is a delta function centered at the frequency of the light. In the more realistic case of broadband light, the interferogram has an oscillating shape, sharply peaked at the zero-path-difference (ZPD) position and the related spectrum is a superposition of all the radiation frequencies, as can be seen in Fig. 2.1b.

The main advantages of the FT-IR Spectroscopy in respect to other techniques that use different spectrometer, like prisms or gratings, are:

- its wide spectral range of investigation is powerful for all the low-energy excitations in solid state physics.
- its very high resolving power, proportional to the length δ of the arm of the scanning mirror.
- its optical precision in determining the optical path difference δ: the position of the scanning mirror is indeed measured by means of a laser light (*Connes advantage*).
- its capability to measure simultaneously all the frequency spectrum, reducing the collecting time by a factor n (where n is the number of points in the spectrum) and improving the signal-to-noise ratio by a factor $N^{1/2}$, where N isa the number of acquired and averaged spectra (*Fellget advantage*).
- its use of a circular aperture that allows one to enhance the light intensity with respect to the case of linear slits (*Jacquinot advantage*).

A disadvantage of this technique is the need of Kramers-Kronig transformations (non local procedure) to extract both the real and imaginary parts of the response functions for bulk systems, requiring extrapolated reflectivity data from zero to the lowest measured frequency and from the highest measured to ∞. This can be avoided by performing two independent measurements, as it is done in ellipsometry or in combined reflectivity and transmittance measurements, to obtain bot the real and the imaginary part of the optical responses by a local procedure.

In the present work, FT-IR has been used to study the response of four bulk samples and two thin films of Topological Insulators. This can be done with FT-IR by sending the radiation modulated by the interferometer onto the sample and then measuring the intensity of either the reflected or transmitted radiation relative to a reference. The resulting reflectivity and the transmittance are

$$R(\omega) = \frac{I_r(\omega)}{I_0(\omega)} \tag{2.2}$$

$$T(\omega) = \frac{I_t(\omega)}{I_0(\omega)} \tag{2.3}$$

where $I_0(\omega)$ is the intensity reflected by a reference mirror or that or transmitted by an aperture. Using our experimental setup, we were able to measure $R(\omega)$ at near-normal incidence in a wide frequency range (from 40 cm^{-1}, to 22,500 cm^{-1}, i.e. from 5 meV to 2.8 eV) and $T(\omega)$ at near-normal incidence in the far-infrared (from 40 cm^{-1}, to 400 cm^{-1}, i.e., from 5 meV to 50 meV) by varying the temperature of the sample from 10 K to 500 K. The frequency of the radiation will be hereafter expressed in cm^{-1} with an implicit factor 2πc. An other disadvantage of the FT-IR technique is the lack of high-flux broadband conventional radiation sources . For that reason it is useful to use a source brilliant even at low frequencies, namely in the TeraHertz (THz) and sub-THz regions (1THz $\cong$ 33cm^{-1}), where the electrodynamics of the TI is concentrated.

2.1.1 Coherent Synchrotron Radiation at BESSY

Synchrotron Radiation (SR) is the radiation emitted by an accelerated charge at relativistic speed with typical energy of the order of a few GeV. SR is a very powerful source for investigations in the ultraviolet and X-ray spectral ranges. Infrared spectroscopy usually uses either black-body sources, or synchrotron radiation. Both these sources are incoherent if we consider the phase relationship in time and space between two different points of the wave front. Indeed, one has

- spatial coherence if two points belonging to a wave front have a fixed phase relation
- temporal coherence if two points along the wave propagation direction and belonging to two different wave fronts have a fixed phase relationship.

However, especially in the far infrared, infrared synchrotron radiation (IRSR) presents several advantages with respect to black bodies. IRSR is a continuous-wave incoherent radiation

- with a strong reduced thermal noise.
- with an intensity over small areas several order of magnitude bigger than those of conventional sources.
- with an emission strongly collimated.
- with a broad-band emission.
- with high degree of polarization.

In a storage ring or a synchrotron, bunches of electrons (or positrons) with mass m e charge q, initially accelerated by a linear accelerator (Linac), circulate along closed paths, recovering their energy when passing through a radiofrequency cavity with an electric field oscillating at MHz frequencies. In Fig. 2.2 it is shown how magnetic dipoles deflect the trajectory of the bunch with a curvature radius ρ. SR is emitted to the observer O along the tangent to the path and collimated within an angle $1/\gamma$ (where $\gamma = \frac{1}{\sqrt{1-\beta^2}}$ is the relativistic factor). In the infrared, the vertical divergence increases with the wavelength as $\lambda^{1/3}$.

Short bunches come out from the radiofrequency cavity and move along the ring, emitting SR pulses in the undulators and in the bending magnets. Each bunch emits incoherent SR with a total intensity equal to the sum of the intensities emitted by all the electrons, irrespective of their relative phase. Hence, the total intensity is proportional to the number of particles and to the fourth power of their energy.

An impulse of duration $\tau/2$ gives an intensity, obtained from the Fourier transform of the impulse, which is

$$I \propto (\frac{\sin(\omega\tau/2)}{\omega\tau/2})^2 = sinc^2(\omega\tau/2). \quad (2.4)$$

It has a maximum frequency ω_{max} due to relativistc effects and given by the equation

$$\omega_{max} \sim \frac{1}{\tau} \simeq 2\frac{\gamma^3 c}{\rho}. \quad (2.5)$$

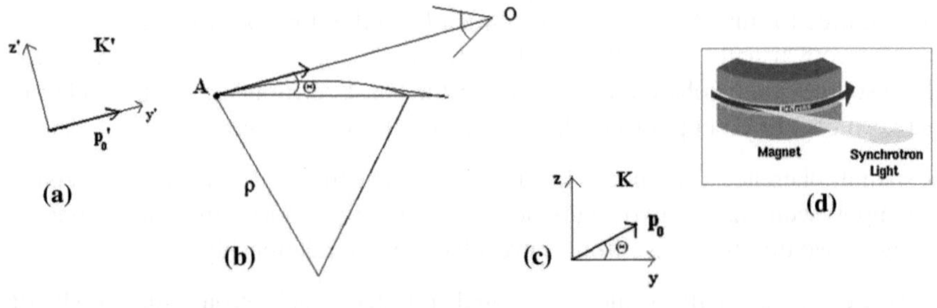

Fig. 2.2 Scheme for emission of Synchrotron Radiation: O is the observer and ρ is the curvature radius (**b**); **K** and **K**$'$ are the reference and laboratory system, respectively, where $\mathbf{p_0}$ ($\mathbf{p}_0'$) is the impulse of the bunch (**a**), (**c**). Sketch for collimated emitted radiation (**d**)

Below 30 cm^{-1} (about 1 THz) the brilliance can be widely increased by using Coherent SR (CSR).

CSR is due to the interference between the electric fields of longitudinally-packed electrons. When the wavelength λ of emitted radiation is comparable with the length of the bunch, all the electrons emit radiation in phase and the irradiated power is proportional to the square of their number (N^2). The emitted power, P, can be derived from the incoherent single-particle power, $P_{1,incoh}$ [1], by the equation

$$P = NP_{1,incoh}(1 + Nf_\lambda) \tag{2.6}$$

where f_λ is a form factor, derived from the Fourier transform of the longitudinal electron density in the bunch. For $Nf_\lambda >> 1$ phase correlation is achieved and mostly coherent synchrotron radiation is emitted with power $\propto N^2$. In the case of a Gaussian bunch-density distribution the form factor is given by $f_\lambda = exp[-(2\pi\sigma/\lambda)^2]$.

The emission of coherent radiation in the FIR (wavelength of few mm) can be obtained for a special magnetic optics operation in third-generation storage rings, which is called low-α mode. In standard synchrotron runs, coherence is suppressed by shielding effects of the dipole vacuum chamber [2, 3]. To overcome this limitation the bunch length and shape are manipulated by tuning the storage ring optics into a dedicated low-α mode, where α is the momentum compaction factor, describing the orbit length variation with beam energy ($\alpha = (\delta L/L)/(\delta p/p)$) [4, 5]. This yields shorter bunches and at higher beam currents of non-Gaussian shape, shifting the CSR spectrum with respect to the shielding cutoff. Above a certain threshold current, bunch instabilities are involved in the emission process, which leads to a periodic or even stochastic bursting [6]. These instabilities can limit the usability of the radiation for spectroscopic applications but enhance the emitted CSR power.

The above described operation mode is available twice a year at the Synchrotron Bessy II. Figure 2.3 shows a comparison between incoherent IRSR and CSR: the

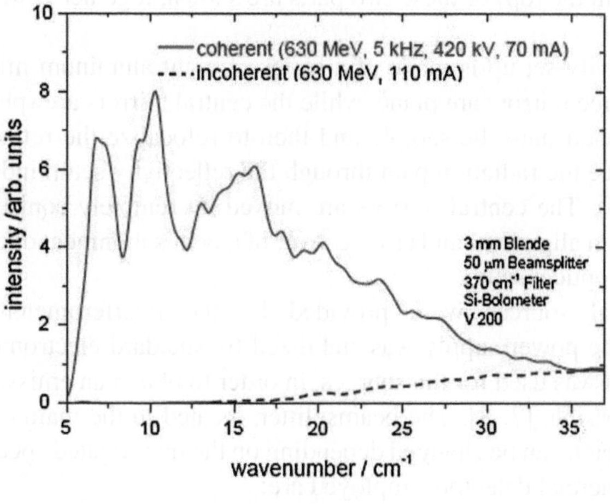

Fig. 2.3 Comparison between incoherent and coherent SR in the sub-THz region at Bessy II

latter one is clearly more intense than the former one in the sub-THz region, with a cut-off at the lowest frequency due to both to the efficiency of the optical components used in experiments and to the losses for diffraction along the optical path.

2.2 Experimental Apparatus

In this section we describe the two experimental apparatus that were used to perform temperature-dependent measurements. The BOMEM DA-3 Interferometer used to measure the reflectivity from mid-infrared to visible of the four TI crystals at IRS laboratory in Rome and the Bruker IFS 66v/S interferometer, used to measure the far-infrared and the sub-THz region of the same sample and the two TI thin films at the beamline IRIS at the Synchrotron BESSY in Berlin. The second apparatus is also the same used at IRS laboratory to measure the patterned TI films. In this section we describe the single parts coupled to the main compartment of the interferometers necessary to measure at low temperature and low frequencies.

2.2.1 *BOMEM DA-3 Interferometer*

BOMEM DA-3 is a vertical interferometer (see Fig. 2.4) whose main compartment allocates sources, beamsplitter and the optics. The moving (scanning) mirror moves within a vertical arm, returning to the initial position with a controlled *flyback*. A horizontal sample compartment is connected to the main compartment and both a close-cycle cryostat and a heating system are inserted into it. The different detectors can be mounted on the modular sample compartment. The two compartments can be evacuated independently to a minimum pressure of about 1 mbar. In Fig. 2.5 the sections (seen from the top) of these two parts are shown, together with the reflectivity setup.

The reflectivity set up is made of a series of eight aluminum mirrors. The first one and last three mirrors are plane, while the central mirrors are spherical, in order to focus the beam onto the sample and then to refocalize the reflected light onto the detector (see the radiation path through the reflectivity setup indicated by a red line in Fig. 2.6). The central mirrors are moved by remotely controlled motors to allow in-vacuum alignment and easy recover of any misalignment due to mechanical stresses of the optical setup.

Conventional sources were provided by the interferometer manufacturer (BOMEM). The power supply was stabilized by standard electronics and a water cooling system was used for the sources, in order to obtain an emission of radiation as stable as possible [7, 8]. The beamsplitter, located in the main compartment of the interferometer, can be changed depending on the investigated spectral range. The standard commercial detectors employed are:

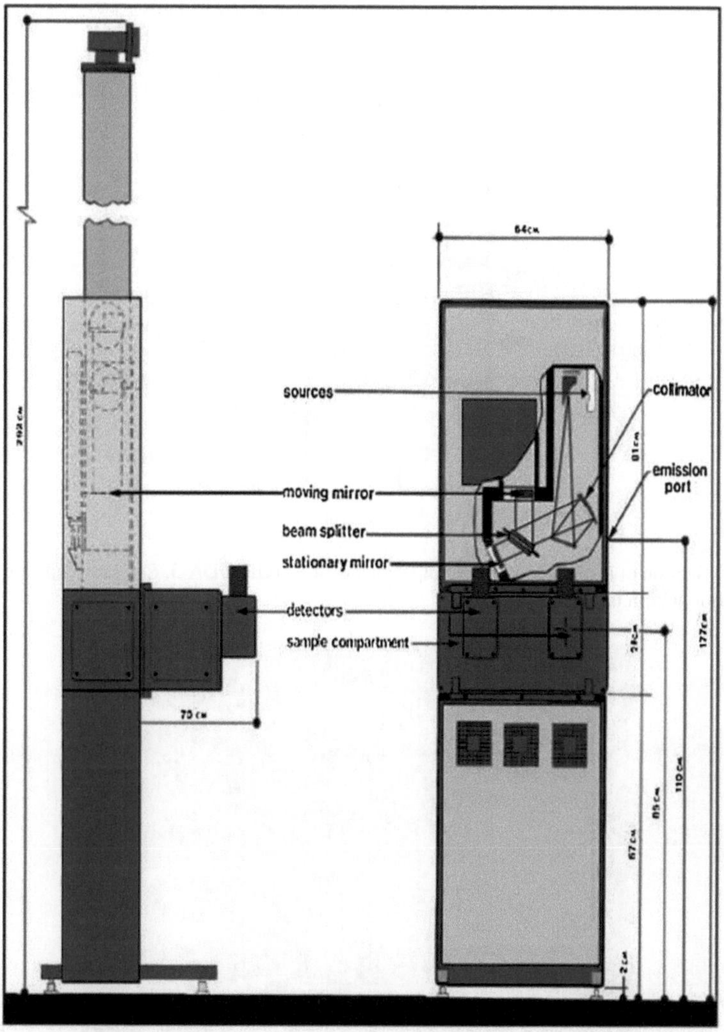

Fig. 2.4 Scheme of BOMEM DA-3: side and front view. The location of sources and beamsplitter are shown, as well as the moving mirror

- *Ge InfraredLabs Bolometer* contained in a Dewar flask and provided of low-passing filters; it works at a temperature of 4.2 K and it is used for a frequency range below 650 cm^{-1};
- a *photoconductive* nitrogen-cooled *HgCdTe detector* used from 500 to 10,000 cm^{-1};
- a *Si photodiode* for frequencies higher than 9,000 cm^{-1}.

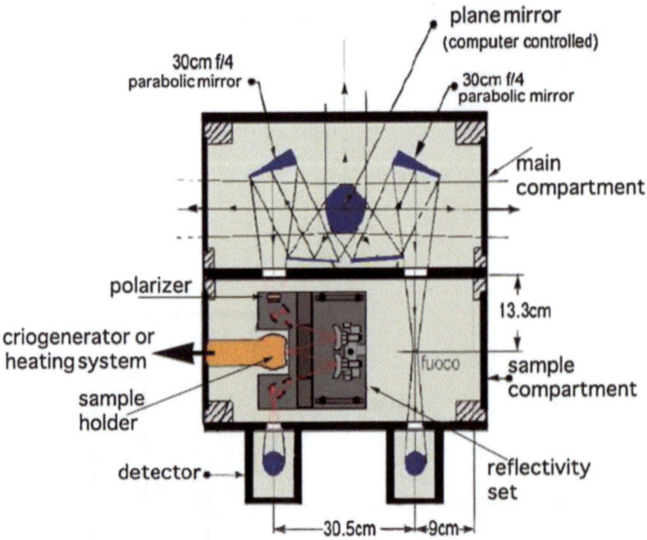

Fig. 2.5 Section of sample and main compartments of BOMEM DA-3. *Red lines* indicate the path of radiation within the reflectivity setup

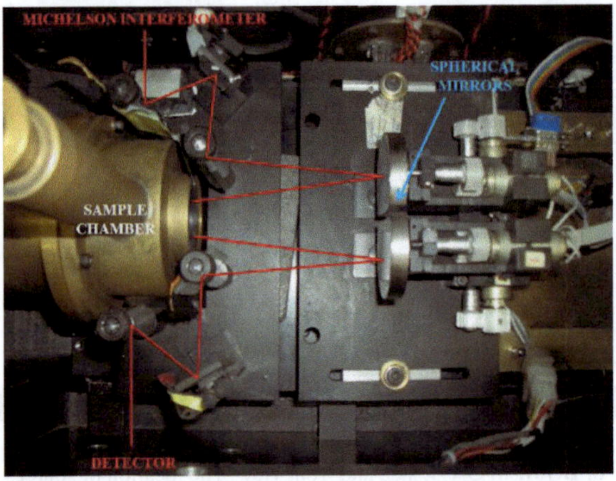

Fig. 2.6 Picture of the reflectivity setup in the sample compartment of BOMEM DA-3. The optical path from the main compartment to the detector is indicated by a *red line*

One can note that the detector ranges do not coincide neither with the frequency ranges of the sources nor with those of the beamsplitters. This means that many measuring sessions were required on the same sample. In Table 2.1, the minimum measuring sessions required to complete the full range, from the far infrared to the visible, for a given sample are listed [9].

Table 2.1 Experimental setup for each of the infrared ranges investigated, defined by the intervals $\omega_L \div \omega_H$

Range	$\omega_L \div \omega_H$ (cm^{-1})	Source	Beam splitter	Polarizer/ window	Detector (T)
far-IR	10–60	Hg arc	Mylar 50 μm	Polyethylene	Bolometer 1.6 K
	20–100	Hg arc	Mylar 25 μm	Polyethylene	Bolometer 1.6 K
	30–220	Hg arc	Mylar 12 μm	Polyethylene	Bolometer 4.2 K
	190–660	Globar	Mylar 3 μm	KRS5	Bolometer 4.2 K
	30–660	Hg arc	Si-covered My6 μm	Polyethylene	Bolometer 4.2 K
mid-IR	500–6,000	Globar	Ge on KBr	KRS5	HgCdTe 77 K
near-IR	4,000–12,000	Quartz-halo	Quartz	KRS5	HgCdTe 77 K
Visible	12,000 24,000	Quartz -halo	Quartz	Polaroid/ Quartz	Si 295K

Note that many far-IR ranges overlap to get independent determination of the far-IR reflectivity

2.2.2 Closed-Cycle Cryostat

A closed-cycle cryostat Leybold COOLPACK 6000 (see Fig. 2.7) has been used in order to perform reflectivity measurements at low temperature. It is a cryogenerator based on a thermodinamic cycle (Gifford-Mac Mahon cycle) able to reach a minimum of 10 K on its cold finger, on the end of which the sample is mounted. A thermometer and a heating coil are mounted on the cold finger, allowing one for regulation of the temperature within 1 K. One can also strongly reduce thermal losses within the second stage screwing a thermal shield on the first stage, surrounding the second one.

The sample is glued by silver paint, to ensure thermal contact, on a brass cone, similar to those described by Homes et al. in Ref. [10]. The cone reflects the radiation that doesn't hit the sample in different directions with respect the normal reflection. The cone can also be aligned, using three tilting screws, to make the sample surface perpendicular to the incident radiation. The final alignment of the sample is made *in situ* using a laser beam or the inner visible source of the interferometer, directly, if the sample reflects well in the visible range.

Moreover, to avoid strong condensation of water, carbon dioxide or a thin solid nitrogen layer over the sample surface, the chamber of the cryostat is kept by two *turbomolecular* pumps at a pressure of about 10^{-6} mbar. Often the cold finger itself it is used as a pump, after a first cooling cycle, allowing one to reach pressures of about 10^{-7} mbar and allowing precise T-dependent measurements also at high frequency.

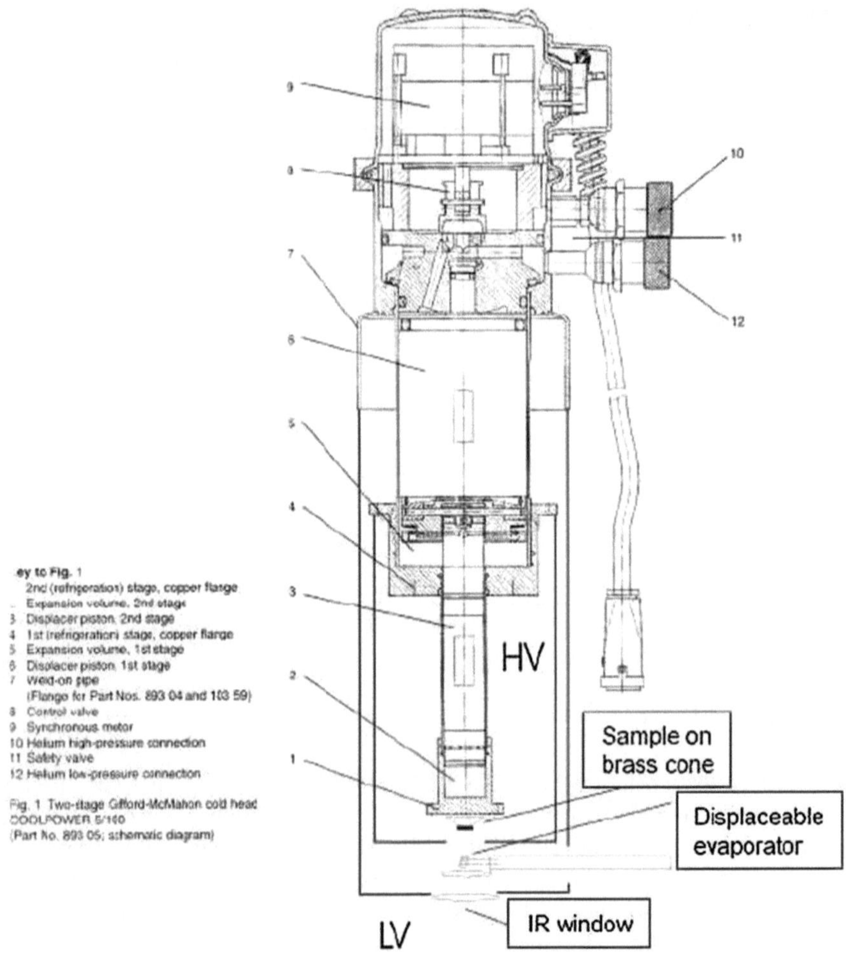

Fig. 2.7 Scheme of the Leybold cryopump with the sample holder on *top*. Two vacuum environments exist: LV (low vacuum, for the interferometer optics) and HV (high vacuum, for the cryostat). An IR window separates the two environments

Infrared spectra have been acquired at the lowest temperature (10 K) with operating cryostat and with the cryostat turned off, re-heating the sample, between the lowest and the room temperature. In the first condition a thermal stability of ± 0.2 K is obtained, and the lowest temperature is reached for an arbitrarily long acquisition time; however, the infrared signal-to-noise ratio is lowered by mechanical vibrations. The re-heating process turned out to be enough slow as to allow spectra acquisition at all the selected temperatures within ± 1 K.

2.2.3 *Infrared Experimental Station at BESSY II: Bruker IFS 66v/S Interferometer*

We have measured the reflectivity and the transmittance at low temperature, down to 5 K, of our samples in the far-infrared and sub-THz regions at the beamline IRIS of the Synchrotron BESSY II at the Helmotz Zentrum in Berlin. A Bruker IFS 66v/S interferometer has been used, shown in a picture in Fig. 2.8a, e in its all parts in Fig. 2.9. It's a versatile instrument, whose components can be all remotely controlled *via* software, so that different measurement configuration can ben set without any need to ventilate the interferometer. The alignment may also be made by using the remote control of any mirror by software. The interferometer is divided into several independent compartments, as shown in Fig. 2.9. The A compartment is the building block of the interferometer, composed by the radiation sources, the beam splitter, the fixed mirror and the mobile mirror capable of moving on an air-cushion; this limits the friction effect on the mirror movement but also can compromise the vacuum conditions in the compartment. In the B compartment finds place the setup for transmittance or reflectivity measurements. The C compartment contains all the internal detector and the E section contains the building block of the electronics of the instrument. As one can see in panel L, the radiation source is in one of the two focal points of a an elliptical mirror. Light coming from the source is driven into the aperture wheels toward a parabolic mirror, which converts it in a plane wavefront. After passing through the beam splitter and covering the optical path in the two arms of the interferometer, it is focused by a second parabolic mirror in the sample compartment (see Fig. 2.8b) and then reaches the detector. The sample is mounted at the end of a vertical liquid-helium cryostat (see below). Some of the key features and performances of the Bruker IFS 66v/s interferometer can be summarized as follows:

- Full spectral range coverage from the Terahertz ($< 5\,\mathrm{cm}^{-1}$) up to the vacuum UV ($\sim 30.000\,\mathrm{cm}^{-1}$).
- Spectral resolving power of better than 100.000 : 1 or $< 0.1\,\mathrm{cm}^{-1}$ spectral resolution.
- Outstanding signal-to-noise: peak-to-peak noise of less than 10-5 AU achieved within 1 min and $4\,\mathrm{cm}^{-1}$ spectral resolution.
- Time resolved spectroscopy: more than 100 spectra/sec Rapid Scan at $12\,\mathrm{cm}^{-1}$ spectral resolution; step scan temporal resolution of < 10 nsec in the mid IR.
- slow Scan with less than 0.006 cm/sec optical velocity.

The different optical configurations used are listed in Table 2.2.

The optical and plasmonic responses of a patterned TI is strongly dependent on the polarization of the electromagnetic radiation. For this reason the measurements of the patterned with a grating TI thin film have been performed with polarized light. A polyethylene-based metallic wire polarizer from 10 to $220\,\mathrm{cm}^{-1}$ with an efficiency of 98 percent has been chosen for the far-infrared, THz ans sub-THz regions, in order to match its range of efficiency with that one of any other window of the optical setup (see Table 2.2).

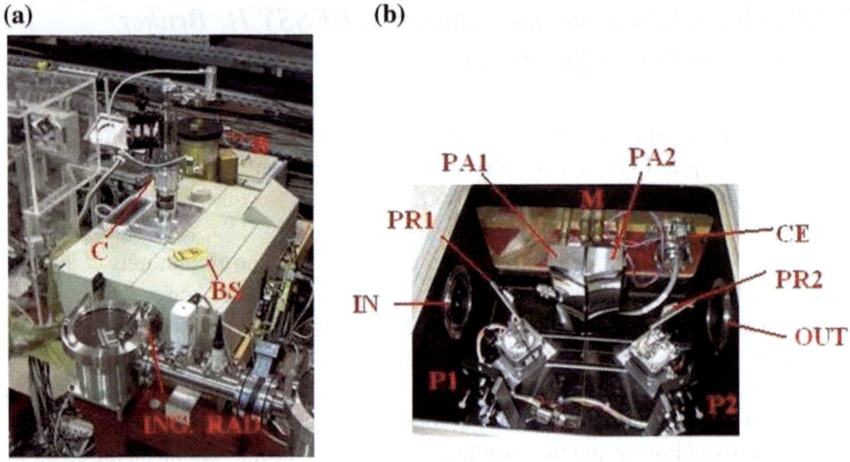

Fig. 2.8 External view of the interferometer with the cryostat mounted on it at BESSY (**a**); reflectivity set inside the Bruker with four plane mirrors PR1, PR2, P1, P2 and two paraboloid mirrors PA1 and PA2. They collect the radiation from IN to the sample and then to the detector (OUT) (**b**)

Table 2.2 Optical configurations used to cover the different spectral range with the Bruker interferometer at BESSY II

Spectral range	Frequencies cm^{-1}	Source	Beamsplitter	Optical window	Polarizer	Detector
FIR	50–600	Globar, Hg	Ge-covered Mylar 6 μm	Polyethylene	Polyethylene	Bolometer 4.2 K
THz	10–100	Hg	Mylar 25–50–125 μm	Picarin	Polyethylene	Bolometer 4.2 K
Sub-THz	5–10	SR	Mylar 50 μm	Picarin	Polyethylene	Bolometer 1.6 K

2.2.4 Liquid-Helium Cryostat and Pumped Bolometer

We used a vertical continuous-flow liquid helium cryostat (*Janis model*) in order to measure low temperature reflectivity and transmittance. It is able to reach a minimum T = 5 K on its coldfinger, at the end of which there is the sample holder, as shown in Fig. 2.10. A thermometer and a heating coil are mounted on the top and around the coldfinger, respectively, in order to ensure an external temperature control within $\pm$ 0.5 K. A thermal shield is screwed on the lower part of the first stage, allowing one to reduce the radiation heating from the surrounding second stage. The vertical under vacuum movement of the sample holder is allowed by a mechanic external stage, which provides micrometric displacements. For transmittance measurements we can mount on the sample holder both the sample and the bare substrate, as reference,

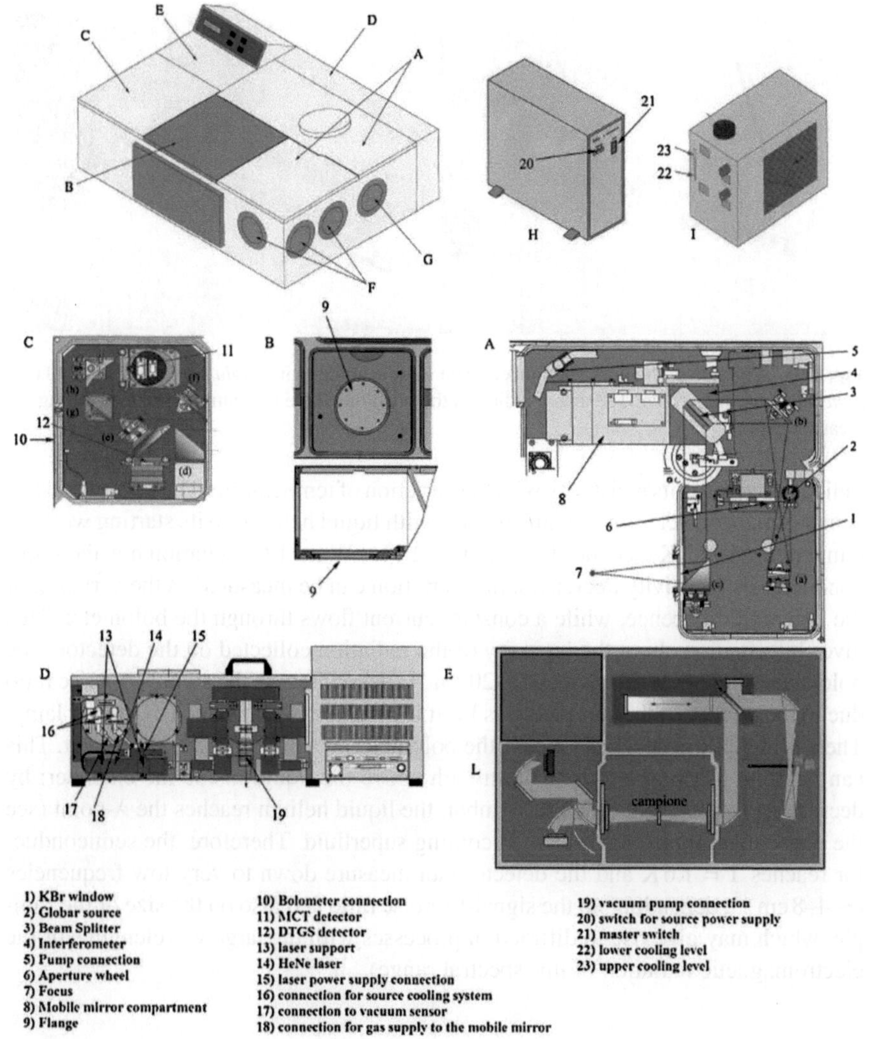

Fig. 2.9 An overall sketch of the Bruker IFS 66v/S interferometer and details of the different components. **A** Michelson's interferometer compartment where a, b, c, d, e, f, g, h indicate the mirror succession, **B** sample compartment, **C** detector compartment, **L** optical scheme of the measurements with the interferometer

in order to measure them at the same time. The samples were mounted by means of brass plates fixed with brass screws, in order to maximize the thermal contact.

In the sub-THz region, using both an inner Hg lamp and the Synchrotron Radiation as a source, we have been used a Ge Bolometer as a detector, which, in particular conditions, can measure down to 4 cm^{-1}. The bolometer is a thermal detector, used in the FIR, made of a doped semiconductor (Ge) which, when hit by electromagnetic

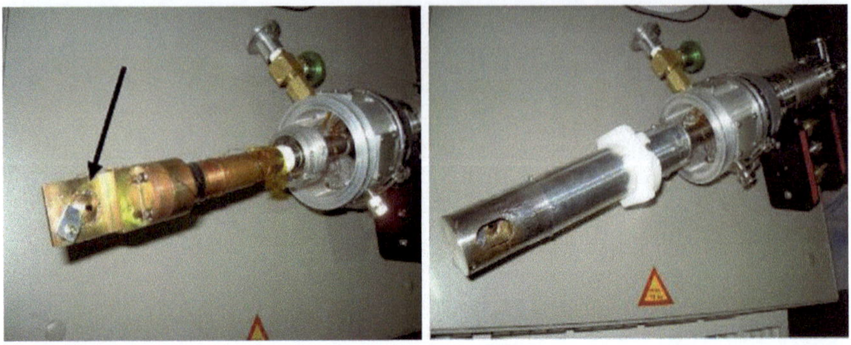

Fig. 2.10 A picture of the liquid helium cryostat sample holder with (*right panel*) and without (*left panel*) the thermal screen. The arrow indicates the position where the sample is placed during the measurements

radiation, changes its resistivity with the variation of temperature. The semiconductor is in thermal contact with a chamber filled with liquid helium, so its starting working temperature is 4.2 K (see the sketch in Fig. 2.11a) When FIR radiation hits the semiconductor its resistivity decreases: this variation can be measured as the variation of the potential difference, while a constant current flows through the bolometer. This gives information about the intensity of the radiation collected on the detector. The bolometer can measure down to 15–20 cm^{-1} (depending on the signal-to-noise ratio due to the quality of the sample): this limit is given by the brilliance of the Hg lamp. Then, with CSR is necessary to use the bolometer in a different working point. This can be done pumping on the helium bath above the liquid inside the chamber: by decreasing the pressure down to 11 mbar, the liquid helium reaches the λ-point (see the phase diagram in Fig. 2.11b), becoming superfluid. Therefore, the semiconductor reaches T = 1.6 K and the detector can measure down to very low frequencies ($\sim$ 4–8 cm^{-1}, depending on the signal-to-noise ratio and also on the size of the sample, which may give rise to diffraction processes with the large wavelength s of the electromagnetic radiation of this spectral range).

2.3 Reflectivity Measurements

The reflectivity measured in this work, at near-normal incidence for frequencies between 30 $\div$ 20,000 cm^{-1}, is defined as the ratio between the electromagnetic intensity reflected from the sample (I_r) and that reflected from a reference mirror (I_0) (see Eq. 2.2).

$R(\omega)$ is related to the complex refractive index $\tilde{n} = n + ik$ via the Fresnel equations [11, 12]:

$$R(\omega) = |\tilde{r}|^2 = |\frac{(\tilde{n} - n_W)}{(\tilde{n} + n_W)}|^2 \tag{2.7}$$

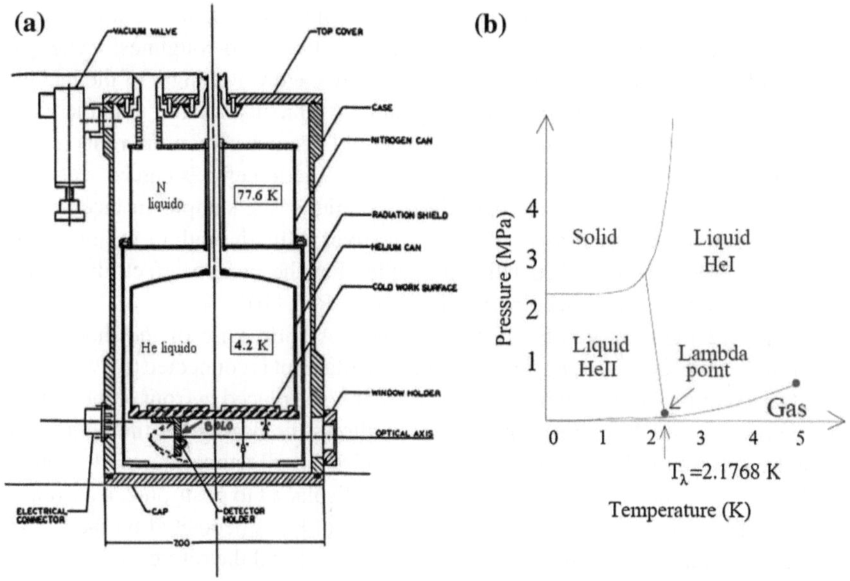

Fig. 2.11 Scketch of the inside part of a bolometer (**a**). Phase diagram of helium near the λ point (**b**)

where r is the reflection coefficient and n_W is the refractive index of the medium that shares the front interface with the sample. Whether the reflectivity is measured at a vacuum-sample interface n_W =1 and Eq. 2.7 reduces to the simple formula:

$$R(\omega) = \frac{(n-1)^2 + k^2}{(n+1)^2 + k^2}. \tag{2.8}$$

Since

$$r(\omega) = ln\sqrt{R(\omega)} + i\theta(\omega) \tag{2.9}$$

where $r(\omega)$ is the complex reflectance, $R(\omega)$ e $\theta(\omega)$ are related by the Kramers-Kronig (KK) transformations, that generally correlate the real and the imaginary part of a linear response function, due to the causality principle (see below).

2.3.1 Measuring the Reference: the Overfilling Technique

In order to obtain the absolute value of the reflectivity (Eq. 2.2), usually both the intensity reflected by the sample and the one reflected from a pre-aligned reference (a mirror) are measured at very temperature.

If, however, the sample is a small and irregular single crystal, diffraction becomes important, because it may arise from the imperfections and roughness, when the surface defect dimension is comparable with the average wavelength of the incident radiation. For commercial mirrors the roughness is typically smaller than the wavelength, therefore different diffraction effects could occur between the reflection by the sample and that one by the reference. Those diffraction effects can be accounted and partially eliminated with the overfilling technique: the sample surface itself is used as the reference, covering it by a gold (or silver) film, through in-vacuum metal evaporation. The thickness of such film is smaller than the surface defects dimension, then the diffraction effects become negligible (see Ref. [10]).

In Fig. 2.12 the evaporator scheme is shown. A short wire of the metal to be evaporated is threaded in a tungsten filament. The filament is connected to a wand, that can be raised or lowered from outside the cryostat, to be placed in front of the sample. The filament is usually in position 1, where it allows the passage of the light beam. When the spectra on the sample have been acquired at all temperatures, the sample is re-heated to room temperature; the filament is displaced to position 2, a current of about 3 A is applied, so that the metal melts and a film is deposited on the sample surface. The filament is then moved back to position 1 and the reference spectra are collected at all temperatures. It is common experience that the best deposition occurs when the sample is at room temperature.

It is worth noting that no optical path is changed during the deposition and that the interferometer is kept evacuated and free of any mechanical stress. Moreover, the overfilling technique allows one to recover the same thermal contractions both on the pure sample and on the metallic deposited foil, performing two different thermal cycles to acquire the spectra.

2.3.2 *Kramers-Kronig Transformations*

It is possible to derive the complex refractive index at every frequency by exploiting the Fresnel relation, equivalent to Eq. 2.7:

$$\tilde{r} = \frac{(\tilde{n} - 1)}{(\tilde{n} + 1} \tag{2.10}$$

If one now uses the Maxwell equations for the electromagnetic field in matter to combine the real and the imaginary part of $\tilde{n} = n + ik = \sqrt{\tilde{\epsilon}}$ [11], one obtains the components of the complex dielectric function $\tilde{\epsilon} = \epsilon_1 + i\epsilon_1$:

$$\epsilon_1(\omega) = n(\omega)^2 - k(\omega)^2 \tag{2.11}$$

$$\epsilon_2(\omega) = 2n(\omega)k(\omega) \tag{2.12}$$

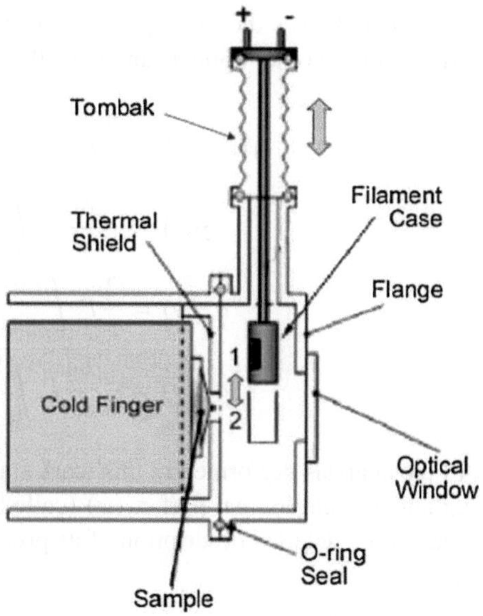

Fig. 2.12 Scheme of the evaporator

Then, using the relation $\tilde{\epsilon} = \epsilon_\infty + 4\pi\tilde{\sigma}/\omega$, where ϵ_∞ is the high-frequency dielectric constant, one can derive the complex conductivity $\tilde{\sigma} = \sigma_1 + i\sigma_2$:

$$\sigma_1(\omega) = \frac{\omega}{4\pi}\epsilon_2(\omega) \tag{2.13}$$

$$\sigma_2(\omega) = \frac{\omega}{4\pi}[\epsilon_\infty - \epsilon_1(\omega)] \tag{2.14}$$

Therefore, one can obtain all the microscopic response function of the system from the only knowledge of the frequency dependent complex reflectance $\tilde{r}(\omega)$. Its imaginary part can be, in fact, evaluated by extrapolating the experimental $R(\omega)$ to $\omega \to 0$ and $\omega \to \infty$ (the free electron asymptotic behavior $R(\omega) \propto \omega^{-4}$ is generally used) and then performing the KK analysis.

KK transformations were introduced by Kramers [13–15] and Kronig [16, 17] in 1926 and still play a fundamental role in condensed matter. These relations can be derived from general considerations involving causality [11, 12] and have a wide application as they allow for the evaluation of the components of the complex dielectric function or the complex conductivity when only one quantity, such as the reflectivity or the absorbed power, is measured.

In the case of $\tilde{r}(\omega)$, the KK relation

$$\theta(\omega) = -\frac{2\omega}{\pi} P \int_0^{+\infty} \frac{\ln\sqrt{R(\omega')}}{(\omega')^2 - \omega^2}\, d\omega' \tag{2.15}$$

where P indicates the principal value of the integral. Similar KK relations can be written for the other optical functions, that is

$$\epsilon_1(\omega) - 1 = \frac{2}{\pi} P \int_0^{+\infty} \frac{\omega' \varepsilon_2(\omega')}{(\omega')^2 - \omega^2} \, d\omega' \tag{2.16}$$

$$\epsilon_2(\omega) = -\frac{2\omega}{\pi} P \int_0^{+\infty} \frac{[\varepsilon_1(\omega') - 1]}{(\omega')^2 - \omega^2} \, d\omega' \tag{2.17}$$

$$\sigma_1(\omega) = \frac{2}{\pi} P \int_0^{+\infty} \frac{\omega' \sigma_2(\omega')}{(\omega')^2 - \omega^2} \, d\omega' \tag{2.18}$$

$$\sigma_2(\omega) = -\frac{2\omega}{\pi} P \int_0^{+\infty} \frac{[\sigma_1(\omega') - 1]}{(\omega')^2 - \omega^2} \, d\omega' \tag{2.19}$$

All the analysis performed in this work start from the complex conductivity and, in particular, from its real part $\sigma_1(\omega)$ (called *the optical conductivity*). Therefore we give a more detailed description of its properties in the next Section.

2.3.3 Optical Conductivity and Sum Rules

The optical conductivity is defined as the linear response function relating the electrical current $\mathbf{J}$ and the excitation due to an electric field $\mathbf{E}$ by the expression $\mathbf{J}(\mathbf{q}, \omega) = \tilde{\sigma}(\omega, \mathbf{q})\mathbf{E}(\mathbf{q}, \omega)$. It is worth noting that in an optical experiment only the transverse conductivity is detected (i.e. ($\mathbf{J} \perp \mathbf{q}$)) due to the transverse properties of the oscillating electrical field, and that excitation of any energy can be detected but only at $|\mathbf{q}| \sim 0$. This latter condition is easily understood considering that, in order to conserve energy and momentum in an absorption process, $\mathbf{q} = \frac{v}{c}\mathbf{k}$ is required, where $\varepsilon = v\mathbf{k}$ is the dispersion relation for the excitation. In condensed matter, no velocity v higher than 10^6 cm/sec (Fermi velocity in metals) is reached, while $c \simeq 3 \cdot 10^{10}$ cm/sec, which implies $|\mathbf{q}| \sim 0$ for any value of ω in the optical range. As it was shown in the previous section, it is possible to obtain, from the measured R(ω), the frequency dependent conductivity $\tilde{\sigma}(\omega)$. We can derive an expression that relates the optical conductivity to microscopic observables that can be easily defined in a N-electron system. The fluctuation-dissipation theorem relates the fluctuations described by a correlation function to the dissipation described by the imaginary part of a susceptibility. In the case of response to the electromagnetic radiation, the interaction Hamiltonian, in the dipole approximation, reads $H_{int} = -\frac{1}{c}\mathbf{J} \cdot \mathbf{A}$, where $\mathbf{A}$ is the perturbing vector potential. The following expression of the fluctuation-dissipation theorem can be derived directly from the Fermi golden rule

$$\sigma_1(\omega) = \sum_n \frac{1}{\hbar\omega V} \int_0^{\infty} dt \langle n | \{\hat{\mathbf{J}}(0), \hat{\mathbf{J}}(t)\} | n \rangle e^{i\omega t}, \tag{2.20}$$

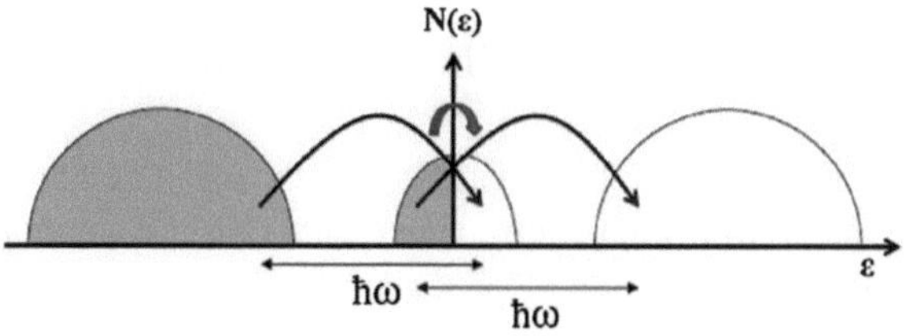

Fig. 2.13 Sketch of a density of states (*grey*: occupied, *white*: unoccupied). The optical conductivity at a given frequency ω is the sum of all the transitions from an occupied to an empty state separated by an energy $\hbar\omega$

which is called the Kubo formula [12].

The right hand side of Eq. 2.20 can be independently derived from the Fermi golden rule for the transition probability between two energy levels, once summed over all the possible initial and final states. This allows one to establish a proportionality relation between σ_1 and the total transition rate W, from which one gets a microscopic interpretation of the conductivity, as the sum over all the possible "jumps" of energy ω in a given distribution of states ε_n, weighted by their dipole matrix element, and whose ground state energy is ε_g (Fig. 2.13):

$$\sigma_1(\omega) = \frac{\pi}{V}\sum_n |\langle n|\hat{\mathbf{J}}|g\rangle|^2 \frac{\delta(\omega - \varepsilon_n + \varepsilon_g)}{\varepsilon_n - \varepsilon_g} \tag{2.21}$$

It can be shown [12] that, by combining Kramers-Kronig relations with physical arguments about the behavior of the real and imaginary part of the response function it is possible to establish a set of so-called sum rules for various optical parameters. We will here define only the *conductivity sum rule* as:

$$\frac{{\omega_p}^2}{8} = \int_0^\infty \sigma_1(\omega, T)d\omega = \frac{\pi N e^2}{2m} = \frac{\pi}{2}\sum_j \frac{q_j^2}{M_j} \tag{2.22}$$

where ω_p is the plasma frequency (defined as $\omega_p = \left(\frac{4\pi N e^2}{m}\right)^{1/2}$). The charge and the mass of the j charged objects in the solid unit cell have been generalized to q and M, respectively, to include the case of phononic excitations. It is worth noting that N is, for more than one electron per atom, the total number of electrons per unit volume if the integral (2.22) is carried out to infinite frequencies. This means that, at high enough frequencies, also the core electrons are excited. The sum rule evaluated up to high frequency expresses a constraint that σ_1 must fulfill when some external parameters vary, such as the temperature T or the pressure P. Once chosen a cutoff frequency Ω to evaluate the integral in Eq. 2.22, the spectral weight is defined by

$$W(\Omega, T) = \int_0^\Omega \sigma_1(\omega, T) d\omega. \tag{2.23}$$

In a conventional material W has a cutoff at ω_p, the plasma edge, which separates low-energy (intraband) from high-energy (interband) charge excitations. The spectral weight is conserved at ω_p .

2.4 Transmittance Measurements

In this work the transmittance of four TI thin films has been measured, both in the case of bare films on Al_2O_3 substrate and in that of patterned films (with a grating of different widths).

As we have already seen, the transmittance in the absence of interference effects, here neglected due to the sample roughness, is given by the Eq. 2.3 or, equivalently, by

$$T(\omega) = \frac{[1 - R(\omega)]^2 e^{-\alpha(\omega)t}}{1 - R(\omega)^2 e^{-2\alpha(\omega)t}} \tag{2.24}$$

where $R(\omega)$ is the reflectivity, $\alpha(\omega)$ is the absorption coefficient and t is the thickness of the sample. Otherwise, the transmittance of a film deposited on a substrate, measured relative to the transmittance of the substrate itself, is related to real and imaginary parts of the sheet conductance $\tilde{\sigma}_\square(\omega)$ of the film [18], by

$$T(\omega) = \frac{1}{\{1 + Z_0\sigma'_\square(\omega)/[\tilde{n}(\omega) + 1]\}^2 + \{Z_0\sigma''_\square(\omega)/[\tilde{n}(\omega) + 1]\}^2} \tag{2.25}$$

Here $Z_0 = \sqrt{\mu_0/\varepsilon_0} = 377\Omega$ is the impedance of free space, $\tilde{n}(\omega)$ is the complex refractive index of the substrate, and $\sigma'_\square(\omega)$ and $\sigma''_\square(\omega)$ are the real part and the imaginary part of the sheet conductance of the film, respectively, where

$$\tilde{\sigma}_\square(\omega) = \tilde{\sigma}(\omega)t \tag{2.26}$$

with $\tilde{\sigma}(\omega)$ the complex conductivity of the film.

2.4.1 Local Procedure for the Extraction of the Conductance

From Eq. 2.25 it is possible to extract directly the real part of the conductance, from which one can derive, by 2.26, the optical conductivity of the material. Hence, inverting the 2.25, we have

$$\sigma'_{\square}(\omega) = \frac{[\tilde{n}(\omega)+1]}{Z_0}\left(\frac{1}{T(\omega)}-1\right) - \sigma''_{\square}(\omega) \tag{2.27}$$

One can calculate the imaginary part of the conductance, fitting the transmittance by the Eq. 2.25, in which one can extract both the real and the imaginary part of the conductance by means of the Drude-Lorentz model (see Sect. 2.6.1).

Actually, in most experiments $\sigma''_{\square}(\omega) \ll [\tilde{n}(\omega)+1]/Z_0$. In this case the contribution of the imaginary part of the conductance to the transmittance is negligible and one can approximate the (2.27) as

$$\sigma'_{\square}(\omega) \simeq \frac{[\tilde{n}(\omega)+1]}{Z_0}\left(\frac{1}{T(\omega)}-1\right) \tag{2.28}$$

Furthermore, if the complex refractive index of the substrate $\tilde{n}(\omega)$=$n(\omega)$+i$k(\omega)$ has the real part n more o less constant as a function of frequency and $k \ll n$, then one can approximates again the (2.28) as

$$\sigma'_{\square}(\omega) \simeq \frac{[n+1]}{Z_0}\left(\frac{1}{T(\omega)}-1\right) \tag{2.29}$$

2.4.2 *Surface Plasmon Polariton*

When an electric field hits a metal, free electrons are displaced from their position. The resulting lack of negative charge gives rise to a Coulombian attraction, which forces them to their original position. This produces an oscillating charge density called *plasmon*, with a characteristic resonance frequency (*plasma resonance*) given by ω_p=$\sqrt{ne^2/\epsilon_0 m}$, where n is the free-carrier density, e and m the electronic charge and mass, respectively, and ϵ_0 the vacuum dielectric constant. One usually refers to this collective mode as *bulk plasmon*. If the charge oscillation is confined at the interface between the metal and a dielectric medium, the relative activated collective mode is called *Surface Plasmon Polariton* (SPPs). Here the term *polariton* is referred to the hybrid nature of collective modes at low wave vectors, when they strongly interact with light [19].

The electromagnetic behavior and the dispersion relation of SPPs can be derived from Maxwell's equations with appropriate boundary conditions in a simple geometric model shown in Fig. 2.14, with the plane $z = 0$ as the interface between the metal and the dielectric medium. The first medium (medium 1, $z < 0$) has permittivity $\epsilon_1(\omega)$ ($\epsilon_2(\omega)$ =0 because no damping is considered), while the second one (medium 2, $z > 0$) has positive permittivity ϵ_d. The direction of propagation of the surface mode is assumed along x, so that the electric field can be written $\mathbf{E}(x, y, z) = \mathbf{E}(z)e^{i\beta x}$, where $\beta = k_x$ is the propagation constant, which corresponds to the component of the wave vector of light along the direction of propagation of the traveling wave.

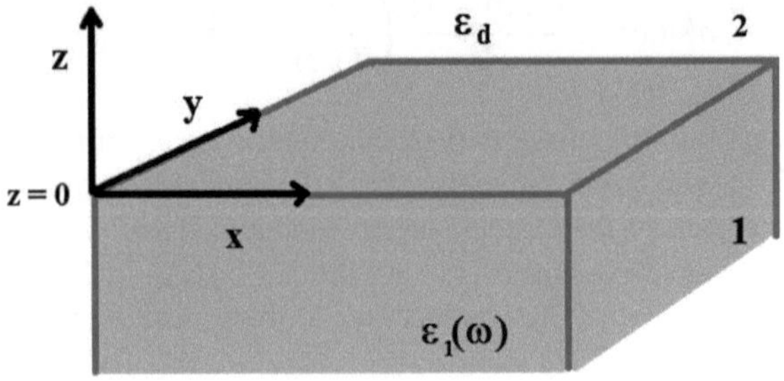

Fig. 2.14 Geometric model for the interface configuration between two mediums, sustaining SPPs

Now, considering a Helmholtz equation where the field has the above form, one can obtain the wave equation for a surface wave, namely

$$\frac{\partial^2 \mathbf{E}(z)}{\partial z^2} + (k_0^2 \epsilon - \beta^2)\mathbf{E} = 0 \tag{2.30}$$

where ϵ is ϵ_1 or ϵ_d for the medium 1 or 2, respectively. If one now uses the Maxwell's equations assuming harmonic time dependence ($\frac{\partial}{\partial t} = 0$), one-dimensional geometry and a uniform field in the plane ($\frac{\partial}{\partial y} = 0$), one has:

$$\frac{\partial E_y}{\partial z} = -i\omega\mu_0 H_x \tag{2.31}$$

$$\frac{\partial E_x}{\partial z} - i\beta E_z = -i\omega\mu_0 H_y \tag{2.32}$$

$$i\beta E_y = i\omega\mu_0 H_z \tag{2.33}$$

$$\frac{\partial H_y}{\partial z} = -i\omega\epsilon_0\epsilon E_x \tag{2.34}$$

$$\frac{\partial H_x}{\partial z} - i\beta H_z = -i\omega\epsilon_0\epsilon E_y \tag{2.35}$$

$$i\beta H_y = -i\omega\epsilon_0\epsilon E_z \tag{2.36}$$

Such system of equations allows for two solutions with different polarization properties: TM (Transverse Magnetic) and TE (Transverse Electric). For the former mode only the components E_x, E_z and H_y are non zero (electric field parallel to the xz plane), while for the latter one only H_x, H_z and E_y are non zero (electric field parallel to the xy plane).

Let us first consider the TM modes. By applying Eqs. 2.37 to the non zero components in both semispaces one obtains:

$$H_y(z) = A_2 e^{i\beta x} e^{-k_2 z} \tag{2.37}$$

$$H_y(z) = A_1 e^{i\beta x} e^{-k_1 z} \tag{2.38}$$

$$E_x(z) = i A_2 \frac{1}{\omega \epsilon_0 \epsilon_2} k_2 e^{i\beta x} e^{k_2 z} \tag{2.39}$$

$$E_x(z) = -i A_1 \frac{1}{\omega \epsilon_0 \epsilon_2} k_1 e^{i\beta x} e^{k_1 z} \tag{2.40}$$

$$E_z(z) = -A_1 \frac{\beta}{\omega \epsilon_0 \epsilon_2} e^{i\beta x} e^{-k_2 z} \tag{2.41}$$

$$E_z(z) = -A_1 \frac{\beta}{\omega \epsilon_0 \epsilon_1} e^{i\beta x} e^{k_1 z} \tag{2.42}$$

where $k_i = k_{z,i} (i = 1, 2)$ is the component of the wave vector perpendicular to the interface in mediums 1 and 2, respectively. The reciprocal value of k_i defines the decay length of the evanescent fields perpendicular to the interface. The requirement of continuity for the electric field at the interface implies

$$A_1 = A_2 \tag{2.43}$$

$$k_2 / k_1 = \epsilon_d / \epsilon_1 \tag{2.44}$$

Equation 2.44 subsists only if the two permittivities are of opposite sign, that is, the confinement implies that surface waves exist only between materials with the real parts of the permittivity of opposite sign. For the magnetic field the continuity brings to

$$k_1^2 = \beta^2 - k_0^2 \epsilon_1 \tag{2.45}$$

$$k_2^2 = \beta^2 - k_0^2 \epsilon_d \tag{2.46}$$

Now, combining Eqs. 2.44 and 2.45 one obtains the dispersion relation of the SPPs propagating at a single interface between a metal and a dielectric, i.e.:

$$\beta = k_{SPP} = k_0 \sqrt{\frac{\epsilon_1 \epsilon_d}{\epsilon_1 + \epsilon_d}} \tag{2.47}$$

where β is real, if no damping factor in the metal electric permittivity has been taken into account.

Considering now the TE mode and repeating the same procedure for the extraction of β, one finds for the continuity of the field that $A_1 = A_2 = 0$, so all the components of the fields are zero: this means that a SPP can exist only as TM mode.

In Fig. 2.15 the dispersion relations of a SPP on a metal/air and metal/silica interface are reported: in that plot SPPs correspond to the part of the dispersion curve on the right side of the corresponding light line of air or silica, due to the bound nature of the SPPs modes. On the contrary, radiative modes, coupled to light, lie at

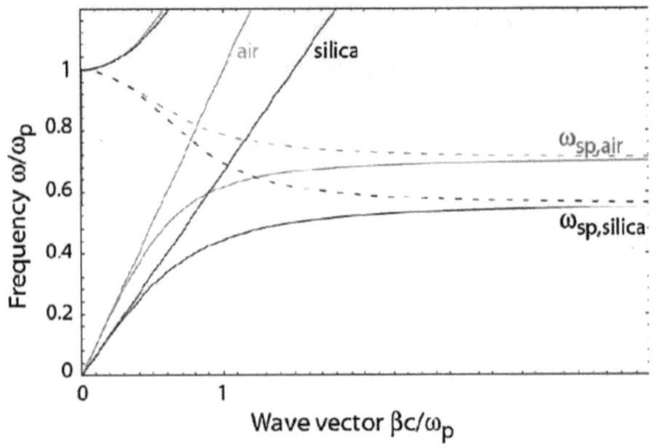

Fig. 2.15 SPP dispersion relations at an interface between metal/air and metal/silica. Both the wave vector and the frequency on the two axis are renormalized with respect to the plasma frequency . Solid curves refer to modes with real propagation constant β, dashed curved refer to non propagating modes (purely imaginary β)

frequencies $\omega > \omega_p$; between the radiative regime and the bound one there is a band gap with purely imaginary wave vectors, where no propagating waves are allowed. In the low-frequency region (wave vectors in the mid-IR range and lower) k_{SPP} is close to the light dispersion line and the SPPs acquire the nature of grazing-incidence light field (the so called Sommerfeld-Zenneck waves [20]). For higher frequencies (larger wave vectors), on the contrary, SPPs approach the surface plasmon frequency $\omega_{SP} = \omega_p/\sqrt{1+\epsilon_d}$: the group velocity of the excited mode goes to zero and the wave (also called *surface plasmon*) acquires an electrostatic character.

As we have assumed so far, no damping has been considered. The metal is treated as an ideal conductor, that is with the imaginary part of the electric permittivity equal to zero. Actually, excitations of the conduction electrons in a metal undergo damping: this leads to a complex electric permittivity and, hence, to a complex propagation constant β. Then, a damped SPP has a finite energy attenuation length L, called propagation length, given by:

$$L = \frac{1}{2Im[\widetilde{\beta}]} \tag{2.48}$$

where $\widetilde{\beta}$ is the complex propagation constant given by Eq. 2.15 in the case of non zero imaginary part of the electric permettivity of the metal. Typical values of the propagation length are in the range $10 \div 100\ \mu$m.

In contrast with the undamped SPP, the damped bound SPP approaches at high wave vectors a maximum finite wave vector in correspondence of ω_{SP}: this fact provides a constraint on the minimum wavelength allowed for propagating SPPs, equal to $\lambda_{SP} = 2\pi/Re[\widetilde{\beta}]$. Moreover this limits also the strength of the field confinement in the direction perpendicular to the interface: the field along z, in fact, decays as $e^{-|k_z||z|}$ with $\widetilde{k}_z = \sqrt{\widetilde{\beta}^2 - \epsilon_d \left(\frac{\omega}{c}\right)^2}$.

Radiation Coupling to a Grating

In the previous Section we have seen that the SPPs dispersion relation lies below the light line: it means that no wave vector matching is possible between a SPP and a photon in that configuration. However, it is possible to excite a SPP introducing a coupling between it and light, just perturbing the periodicity along the metallic surface with a period multiple of the electromagnetic half-wavelength.

One can consider a system with the same periodicity along x and y as shown in Fig. 2.16a. For the wave equation solution a deformation of the surface like that acts as a spatially periodic perturbation with a certain period a, equal to the distance between two following defects (holes or other features). That structure produces a series of reflection planes for the electromagnetic wave, which will be diffracted according to the Bragg condition $n\lambda = 2a$, with n an integer and λ the SPP wavelength. Hence, the periodicity defines a Brillouin zone in the reciprocal lattice along k_x. That mechanism is analogous to the formation of gaps at the Brillouin zone boundary in the quasi-free electron model in crystalline solids and can be easily extended along the other directions when considering a 3D problem.

Therefore, the reciprocal lattice is able to provide the wave vector necessary to excite SPPs through an electromagnetic wave. The resulting conservation of the wave vector is:

$$\mathbf{k}_{SPP} = \mathbf{k}_x \pm i\mathbf{G}_x \pm j\mathbf{G}_y \tag{2.49}$$

where i, j are integer (indicating the order of the scattering that couples the incident wave and the SPP), $\mathbf{k}_x = \mathbf{k}_0$ senθ is the component of the light wave vector parallel to the interface along x and $|\mathbf{G}_x| = 2\pi/a_x$ and $|\mathbf{G}_y| = 2\pi/a_y$ are the reciprocal lattice vectors associated with the two periodicities of the perturbation. For a square array $a_x = a_y = a$, hence $|\mathbf{G}_x| = |\mathbf{G}_y| = 2\pi/a$. From Eqs. 2.49 and 2.47, one can obtain the dependence of the SPPs frequency on the in-plane wave vector on the periodic structure, that is

$$\omega = \left(\frac{\epsilon_1 + \epsilon_d}{\epsilon_1 \epsilon_d}\right)^{1/2} \sqrt{k_x^2 \pm 2i\left(\frac{2\pi}{a}\right)k_x + (i^2 + j^2)\left(\frac{2\pi}{a}\right)^2} \tag{2.50}$$

which can be written at normal incidence as:

$$\omega = \left(\frac{2\pi}{a}\right)\left(\frac{\epsilon_1 + \epsilon_d}{\epsilon_1 \epsilon_d}\right)^{1/2} \sqrt{i^2 + j^2} \tag{2.51}$$

It is worth noting that, when comparing Eqs. 2.49 and 2.47, it has been used an approximation, since the SPP dispersion, appropriate for a smooth film, neglects the fact that the periodic pattern may cause both a significant change in the dispersion and a large coupling between the front and the back surface of the metal film. In fact, as one can see in Fig. 2.16, Eq. 2.47 predicts the position of the plasmonic resonances of the unfolded SPP dispersion and a gap is opened at the edge of the Brillouin zone,

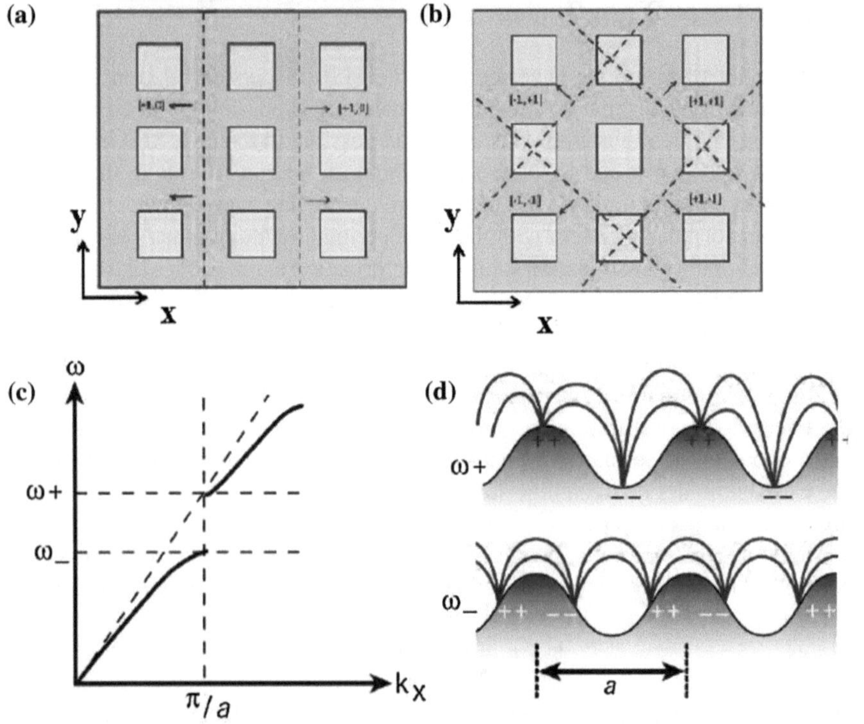

Fig. 2.16 Scheme of the lowest order scattering of SPP excited on a periodic hole array which corresponds to the orders (±1, 0) (**a**) and (±1, ±1) (**b**). The arrows indicate the SPPs propagation direction. Periodic pattern of the metallic interface opens a band gap at a SPP wavelength equal to twice the period at the edges of the Brillouin zone (**c**). This leads to the formation of two standing wave solutions due to the different field and surface charge distribution (**d**) [21]

so that no propagating modes are possible at the frequency predicted by the smooth film equation. Actually, at the band edge the dispersion bands are flat. Therefore the group velocity of the SPP is equal to zero and the SPP density is extremely high. It means that at those frequencies one can find several SPP modes associated with different wave vectors, but to the same energy. Hence, a strong field rises in proximity of the metal surface at those frequencies. The strong enhancement of the electromagnetic field is one of the most important phenomena associated with SPPs.

Such process of localization and consequent enhancement of the field is due to a large propagation constant ($\beta \gg nk_0$, where n is the refractive index of the dielectric at the interface), therefore the exponential decay of the field results in its strong confinement at the interface. The upper limit for β in real metals (with electron damping) is fixed at the wave vector value corresponding to the surface plasma frequency. Due to their high free electron density ($\sim 10^{23}\,\mathrm{cm}^{-3}$), most of common metals have their plasma frequency in the visible or ultraviolet frequency range. Therefore, in metals SPPs are strongly confined at the interface for visible

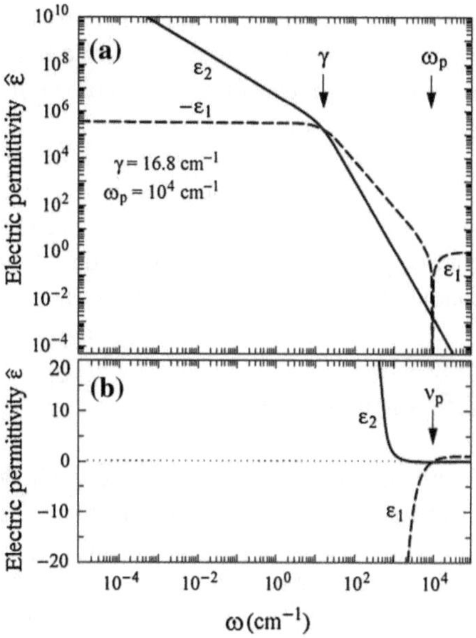

Fig. 2.17 Real and imaginary part of the electric permittivity of a metal with $\omega_p \sim 3 \times 10^{14}$ s^{-1} and $\gamma \sim 0.5 \times 10^{12}$ s^{-1} in logarithmic (**a**) and linear (**b**) scale [12]

frequencies or higher: the so called *plasmonic regime* is established in that range. The requirement on the electric permittivity of the metal for a strong plasmonic regime is that the ratio of its real part over its imaginary part is much larger than unity, *i.e* $|\epsilon_1|/\epsilon_2 \gg 1$. In Fig. 2.17 an example of the real and imaginary parts of the electric permittivity af a metal with $\omega_p \sim 3 \times 10^{14}$ s^{-1} (10^4 cm^{-1}) and damping $\gamma \sim 0.5 \times 10^{12}$ s^{-1} (16 cm^{-1}) is reported. For frequencies up to γ the absolute value of the real part of the electric constant is smaller than its imaginary part: this fact results in a non plasmonic regime ($|\epsilon_1|/\epsilon_2 \ll 1$); on the contrary, above the frequency corresponding to the damping and below the plasma frequency the plasmonic regime is totally achieved.

If the frequency decreases from the visible range down to the THz range, the plasmonic regime weakens. In fact, the absolute value of the real and imaginary parts of the electric permittivity of metals are of the order of about 10^5. This leads to a negligible penetration of the electric field inside the metal, hence to a strong delocalization [19, 22].

Some works showed that at 1 THz the electric field extends up to several centimeters above the metallic interface sustaining SPPs. According to their delocalized nature, SPPs in the THz range behave like a homogeneous field incident under a grazing angle on the metal interface, so they are often called Sommerfeld-Zenneck waves, as already mentioned above [23]. In the limit of Perfect Electric Conductor (PEC) (infinite real part of the electric permettivity), the electric field inside the conductor is equal to zero since no penetration is allowed: this means that a PEC can not support electromagnetic surface modes, like SPPs. However it has been demonstrated in Ref. [24] that a PEC film can sustain bound surface modes, which mimic

SPPs if a periodic perturbation is introduced. The periodic structure, in fact, allows the penetration of an average electric field and restores the conditions required for the existence of a *spoof* SPP, which dominates over the Sommerfeld-Zenneck waves given by the unpatterned conductor.

An expedient to increase the confinement of the SPPs in the THz range is the employment of poor metals or semiconductors, even highly doped: the capability to tune their carrier density, that is the plasma frequency of SPPs, by means of thermal control, photocarrier generation or direct carrier injection makes these materials good candidates for optoelectronic devices.

2.5 Sample Preparation

In this section we will briefly describe how the crystals and the films of topological insulators (TIs), measured for this work, have been grown. Two different techniques have been used by two experimental groups: the modified Bridgeman method for the crystals growth by the Prof. R. J. Cava's group at the Department of Chemistry of the Princeton University (USA) and the Molecular Beam Epitaxy (MBE) for the films deposition by the Prof. S. Oh's group at the Department of Physics and Astronomy of Rutgers, the State University of New Jersey (USA).

2.5.1 Crystal Growth

Four Topological Insulator single crystals have been measured in this work: Bi_2Se_3, $Bi_{2-x}Ca_xSe_3$ with x = 0.0002, Bi_2Se_2Te and Bi_2Te_2Se. They were all grown in the Solid State Chemistry Research Group of Prof. Robert J. Cava at the Princeton University by the modified Bridgman and Bridgman-Stockbarger crystal-growth techniques.

The single crystal of $Bi_{2-x}Ca_xSe_3$ was grown by the "modified Bridgman" method, that is via a process of two-step melting, starting with mixtures of high-purity elements (Bi, 99.999 %; Se 99.999 %; Ca 99.8 %). First, stoichiometric mixtures of Bi and Se were melted in evacuated ampoules at 800 °C for 16 h. Then, the melts were mixed before leading them to solidify by air quenching to room temperature. Second, the stoichiometric amount of Ca was added in the form of pieces, avoiding its contact with the quartz of the ampoule. Therefore, the materials were heated in evacuated quartz ampoules at 400 °C for 16 h and at 800 °C for a day. The crystal growth occurred by cooling from 800 to 550 °C over a period of 24 h, following an annealing at 550 °C for 3 days . Finally, the crystals were furnace cooled to room temperature, ready to be easily cleaved along the basal plane and cut into approximately $1 \times 1 \times 6\,mm^3$ rectangular bar samples [25].

As expected for small-band-gap semiconductors, the quality of the crystals Bi_2Se_2Te and Bi_2Te_2Se from the perspective of the defects and the resultant carrier concentrations is strongly affected by small inhomogeneities in the chemi-

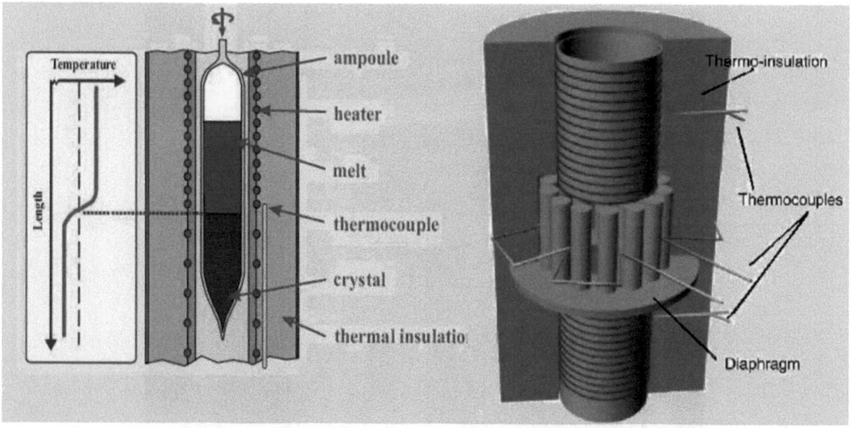

Fig. 2.18 Cross section of the Bridgman-Stockbarger apparatus for the crystal growth (**a**) and three-dimensional view (**b**)

cal composition that occur during the crystal growth. In order to overcome that difficulty, a Bridgman-Stockbarger method was employed to fine-tune the chemical composition of the crystals and optimize the carrier concentration. Actually, single crystals of Bi_2Te_2Se were prepared by two methods: the "modified Bridgman" method and the classical Bridgman-Stockbarger method, whose apparatus is shown in Fig. 2.18. Using the first technique, five-gram mixtures of high purity elemental Bi, Te and Se were sealed in quartz ampoules and then heated up to 850 °C for 1–2 days, followed by cooling to 500 °C at 6–12 °C/h. The samples were then annealed at 500 °C for 3–4 days. The crystals obtained are within a large monolithic piece ($\sim 1\times1\times4\,cm^3$), that usually consists of approximately ten grains presenting random crystal orientations. For the powder XRD (X-Ray Diffraction) characterization of the laboratory-made Bi_2Te_2Se, the samples from the "modified Bridgman" method were in addition annealed at 400 °C for over 2 weeks and then quenched in cold water [26].

Samples of the related compound Bi_2Se_2Te were similarly prepared in according to the similar structure. The second method was also employed for those crystals, allowing for a fine-tuning of the chemical compensation near the stoichiometry composition, being natural variations in it along the directionally solidified crystal boule. Thirty grams of mixture was sealed in a long internally carbon-coated quartz ampoule (20 cm long and 0.8 cm of diameter). This ampoule was tapered at the bottom in order to favor seed selection, and then placed in a vertical furnace (see the picture in Fig. 2.19). The temperature profile of the furnace was set to ensure that the zone hotter than the melting temperature of the Bi_2Te_2Se was longer than the length of the liquid. The temperature gradient at the furnace position crossing the melting point of the Bi_2Te_2Se was $\sim$ 30 °C/cm. The ampoule was then lowered through the hot zone at the speed of 2–4 mm/h. The crystal boules obtained were about 14 cm long with fewer than 10 crystals, which were all grown with their *ab* planes parallel

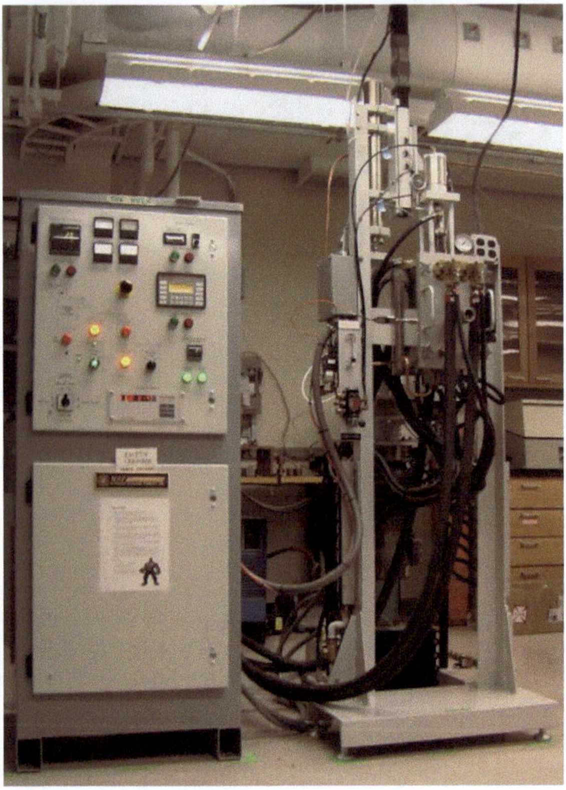

Fig. 2.19 Vacuum Furnace (minimum pressure 10^{-9} Torr, maximum temperature 1,800 °C) for the crystal growth at the Princeton University

to the long axis of the ampoule. The obtained uniform crystal morphology indicates that the boule is relatively homogeneous on a large scale, because its chemical composition gradually varies along the long axis during the directional solidification. The boule was finally cut to seven pieces of about 2 cm equal length, ready for transport measurements.

2.5.2 Film Deposition

The films of topological insulator Bi_2Se_3 of two different thicknesses, here measured, were grown at the Rutgers University by a Molecular Beam Epitaxy (MBE) technique on 0.5 mm thick sapphire substrates Al_2O_3 [27]. A custom-designed SVTA MOS-V-2 MBE system, whose base pressure was lower than 5×10^{-10} Torr, was used. Bi and Se fluxes were provided from Knudsen cells: the fluxes were measured using a quartz crystal microbalance Inficon BDS-250, XTC/3.

In order to start with a clean substrate surface, Al_2O_3 (001) has been exposed to an ex situ UV ozone cleaning step before mounting it in the growth chamber to burn off most organic compounds that may be present on the surface. Then, to remove any further eventual contaminants from the substrate surface, the sapphire was heated to 700 °C in oxygen pressure of 10^{-6} Torr for 10 min. In order to monitor the cleaning and the growth of the samples, the substrate was observed with RHEED (Reflection High-Energy Electron Diffraction) before and after the treatment, by which a bright specular spot and Kikuchi lines were observed after heating and then cooling the substrate [28]: Fig. 2.21a, b indicates that this procedure improved the surface conditions. Therefore, using the two-temperature growth process, Bi_2Se_3 films of various thickness were grown: their surface evolution during the growth was monitored by RHEED, as shown in Fig. 2.21c–f. After deposition of 3 quintuple layers (QLs) (1 QL[1] is about 1 nm) of Bi_2Se_3 at 110 °C, a sharp streaky pattern was observed, providing evidence for a single-crystal Bi_2Se_3 structure. Then the film was slowly annealed to a temperature of 220 °C, in order to further help the crystallization of the film, as one could see by the brightening of the specular spot. The diffraction pattern and the Kikuchi lines became increasingly sharp upon further Bi_2Se_3 deposition: this marks that the grown films have atomically flat morphology and high crystallinity [28]. The film quality was further improved by annealing the sample at 220 °C for an hour after the growth, leading to high quality single crystalline films with large planes and minimal bulk conduction [28].

In Fig. 2.20a a sketch of the MBE apparatus is shown. Relatively precise beams of molecules (heated up so they're in gas form) are shot at the substrate from "guns" called effusion cells. One needs one "gun" for each different beam, shooting a different kind of molecule at the substrate, depending on the nature of the crystal. The molecules land on the surface of the substrate, condense, and build up very slowly and systematically in ultra-thin layers, so the complex, single crystal grows one atomic layer at a time. Separate beams fire different molecules and they build up on the surface of the substrate, arranging epitaxially on top of it. Figure 2.20b shows the nutshell of the MBE apparatus.

2.5.3 Patterning the Thin Films for Plasmonic Studies

The fabrication of the patterned TI thin films by a grating was performed in collaboration with the Institute for Photonics and Nanotechnolgy (IFN) in Rome. The IFN facility has a class 100–1000 Clean Room, that is an environment with a low level of pollutants (hundreds of particles/cubic feet having a 0.5 μm dimension). The Clean Room, 300 square meters wide, is equipped with several devices for film deposition,

[1] TIs have a layered structure with five atomic layers as a basic unit, named a quintuple layer (QL), and the crystal structure is formed by the relatively strong covalent bond within a QL and the weak van der Waals interaction between QLs.

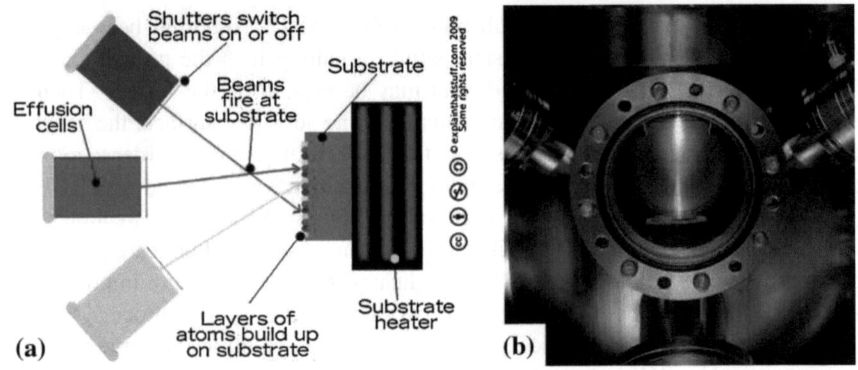

Fig. 2.20 **a** Sketch of Molecular beam epitaxy (MBE) apparatus, creating a single crystal by building up orderly layers of molecules on top of a substrate (base layer). **b** Molecular beam epitaxy (MBE) in action. MBE takes place in ultrahigh vacuum chambers like this, at temperatures of around 500 °C, to ensure a totally clean, dust-free environment; the slightest contamination might damage the crystal. Photo by Jim Yost, courtesy of US DOE/NREL (U.S. Department of Energy/National Renewable Energy Laboratory)

electronic and optical lithography, etching processes and diagnostic. In the following two sections we will shortly describe the fabrication technique of our optical devices.

2.5.4 Electron Beam Lithography

The fabrication of optical devices from Topological Insulator thin films, with thickness of about one hundred of nanometers, requires techniques capable of structuring material on a fine scale. Depending on the shape of the pattern to be fabricated, different resolutions are required, and therefore different techniques must be considered.

The most used fabrication technique for industrial application is Optical Lithography. It consists in the transfer of a pattern to a photosensitive material by selective exposure to a radiation source such as light. A photosensitive material is a material that experiences a change in its physical properties when exposed to a radiation source. If the exposure is selective, that is by masking some of the radiation with a resist film, a pattern on the material exposed is transferred, providing a difference between the properties of the exposed and unexposed regions. It is worth noting that the resolution of this technique is limited by the wavelength of the light source used to expose the resist.

For scientific application a higher resolution for the optical devices is given by the Electron Beam Lithography (EBL). EBL is a process similar to the Optical Lithography, except for using an electron beam instead of light for the exposure. This technique was first introduced in the sixties of last century using Electron Scan

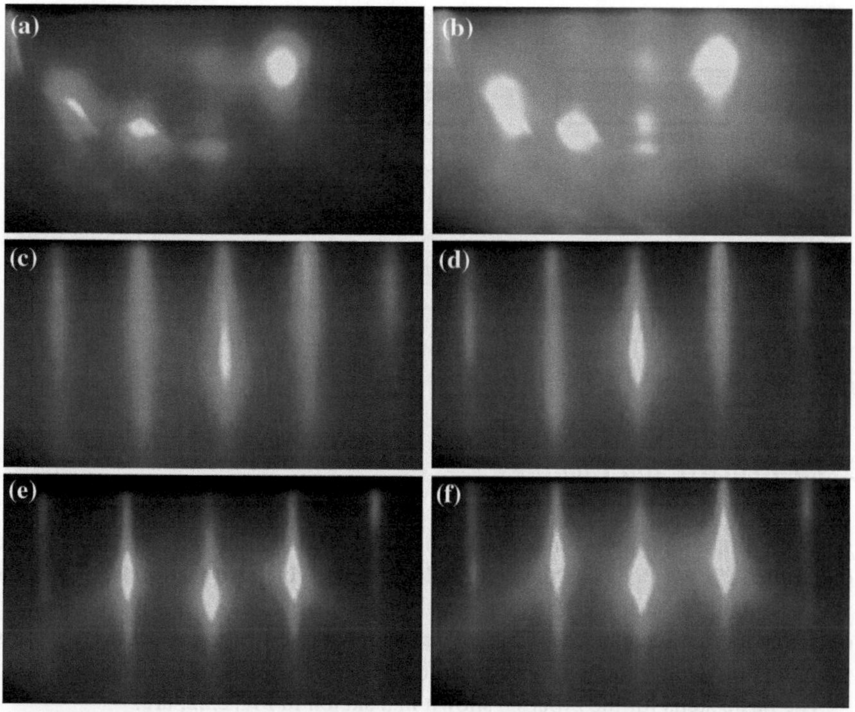

Fig. 2.21 RHEED (Reflection High-Energy Electron Diffraction) images showing the steps of Bi_2Se_3 growth on sapphire substrates. Sapphire substrate mounted in the UHV growth chamber after UV-cleaned for 5 min (**a**). On heating to 700 °Cin an O_2 pressure of 1x 10^{-6} Torr for 10 min (**b**). After deposition of 3 QL of Bi2Se3 film at 110 °C (**c**). Specular beam spot gets brighter on annealing the film to 220 °C (**d**). RHEED pattern gets much brighter and sharper on subsequent growth of another 29 QL at 220 °C (**e**). Final RHEED pattern of the 32 QL film after being annealed at 220 °C for an hour (**f**) [28]

Microscopes (SEM). Its key advantage consists in the possibility to overcome the limit of light diffraction, allowing for nanometric devices fabrication. Moreover the exposure can be done in batch processing. The main disadvantage is the long time that the exposure takes.

The EBL machine has an electron gun able to generate an electron beam with tunable current. The gun is made of a thermoionic emission cathode and uses some electrostatic and magnetic lenses to focus the electron beam up to a spot size of 2–4 nm on the sample. Usually the electron beam is sent on a substrate covered with an electronic resist and is deflected in order to write directly the pattern on the substrate. It has been used the EBL machine *Leica Microsystem EBPG 5000* at IFN for the fabrication of our samples.

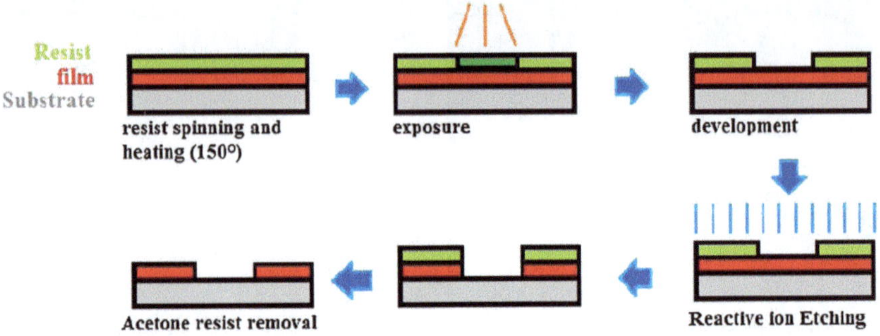

Fig. 2.22 Steps of the Etching procedure

2.5.5 *Reactive Ion Etching*

After the lithography procedure, one can use two fabrication process: the additive one, called *lift-off*, that adds the layer to be patterned on a bare substrate, a process very common for metallic films; the second one, that we have used, is the *etching*, common for semiconductors. The *etching* is a subtractive procedure which starts from the film deposited on a substrate, and patterns it by means of a chemical agent.

The basic steps for the *process*, shown in a sketch in Fig. 2.22, are:

- Cleaning: the sample is initially heated to a temperature sufficient to drive off any moisture that may be present on the surface.
- Coating: the sample is covered with photoresist by *spin coating*: a viscous, liquid solution of photoresist is spread over the sample surface and it is spun rapidly to produce a uniformly thick layer.
- Baking: the photoresist-coated sample is pre-baked, typically at 90–100 °C for 30–60 s on a hotplate, to drive off the excess photoresist solvent.
- Exposure: the EBL process writes the desired pattern on the sample. When exposed, positive photoresist becomes soluble in the basic developer, while the negative one becomes insoluble in the organic developer.
- Development: the chemical change in polymer bonds allows some of the photoresist to be removed by a solution called developer.
- Reactive Ion Etching (RIE): a chemically reactive plasma removes the material deposited on the substrate, which isn't yet covered by the resist. The plasma is generated under low pressure by an electromagnetic field. High-energy ions of the plasma attack the wafer surface and react with it.

Instead of using the RIE, which is a dry process, one can utilize a *wet etching*: it consists of an immersion of the sample in a chemical solution, that removes the uppermost layer not protected by the resist mask. In Fig. 2.23 images at the Electronic Microscope of the four TI patterned films are shown.

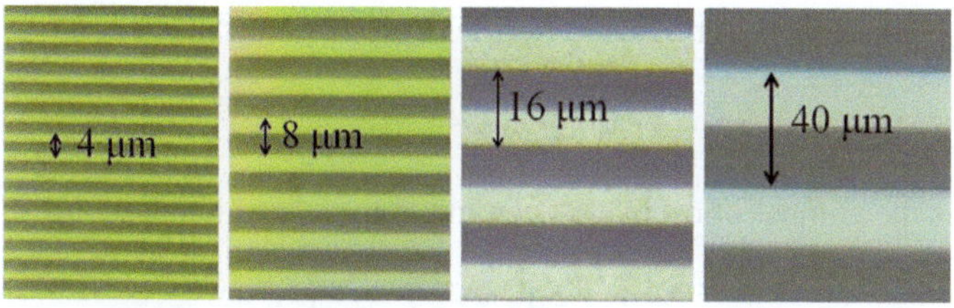

Fig. 2.23 Images at the Electronic Microscope of the four TI films (of different width of grating) after the EBL and RIE processes

2.6 Data Analysis and Fitting Models

2.6.1 The Drude-Lorentz Model

The optical conductivity of a metallic or insulating system can be decomposed into a sum of contributions related to different charge and lattice excitations. The Drude model derives from a classical analysis of the transport properties of a Fermi liquid, where the quasi-elastic scattering processes are controlled by one characteristic time. This leads to an optical conductivity of the form

$$\sigma_1(\omega) = \frac{\omega_p^2}{4\pi}\frac{\tau}{1+(\omega\tau)^2} \tag{2.52}$$

where $\omega_p = \sqrt{\frac{4\pi ne^2}{m}}$ is the plasma frequency of the Fermi liquid, n is the number of charge carriers for volume unit and m is the effective mass.

The Lorentz model, on the contrary, classically describes an insulator, where the atom can be modeled as a valence electron bound to a nucleus. The equation of motion of such a system is that of a driven damped harmonic oscillator, in which the driving force is due to the local electric field, while the damping is due to a dissipative term proportional to the velocity:

$$m\frac{d^2\mathbf{r}}{dt^2} + m\gamma\frac{d\mathbf{r}}{dt} + m\omega_0^2\mathbf{r} = -e\mathbf{E}_{loc} \tag{2.53}$$

In the linear response regime, the polarizability of the system $\tilde{\alpha}(\omega)$ is

$$\mathbf{p} = \tilde{\alpha}(\omega)\mathbf{E}_{loc} = -e\mathbf{r} \tag{2.54}$$

where $\mathbf{p}$ is the electric dipole. If one solves the Eq. 2.53, one obtains

$$\tilde{\alpha}(\omega) = \frac{e^2/m}{\omega_0^2 - \omega^2 - i\gamma\omega}. \tag{2.55}$$

As the polarizability is related to the dielectric function $\tilde{\epsilon}(\omega)$ by

$$\tilde{\epsilon}(\omega) = 1 + 4\pi N \tilde{\alpha}(\omega)$$

where N is the number of oscillators per unit volume, one obtains

$$\tilde{\varepsilon}(\omega) = \varepsilon_{\infty} + \frac{4\pi N e^2}{m} \frac{1}{\omega_0^2 - \omega^2 - i\gamma\omega} \tag{2.56}$$

with ε_{∞} equal to the dielectric function at high frequencies.

If the N oscillators have widths γ_j and resonance frequencies ω_j, one can write

$$\tilde{\varepsilon}(\omega) = \varepsilon_{\infty} + \sum_{j=1}^{N} \frac{4\pi N_j e^2}{m} \frac{1}{\omega_{0j}^2 - \omega^2 - i\gamma_j\omega} \tag{2.57}$$

In Eq. 2.57, the contribution of free carriers, the so called Drude term, is obtained setting $\omega_{0j} = 0$, considering ω_{0j} as the frequency of the transition from the electronic ground state to the excited state. Hence, the Drude term is

$$\tilde{\varepsilon}(\omega) = 1 - \frac{\omega_p^2}{\omega^2 + i\Gamma_D\omega} = 1 - \frac{\tau\omega_p^2}{\tau\omega^2 + i\omega} \tag{2.58}$$

where $\omega_p^2 = \frac{4\pi n e^2}{m}$ is the plasma frequency of free electrons and $\tau = \frac{1}{\Gamma_D}$ is the average time between two electron-phonon, electron-impurity, electron-electron scattering provided by the Matthiessen rule [29]:

$$\frac{1}{\tau} = \frac{1}{\tau_{e-ph}} + \frac{1}{\tau_{e-e}} + \frac{1}{\tau_{e-imp}}. \tag{2.59}$$

If the system is a crystal, one has to add to 2.57 and 2.58 (Drude-Lorentz model) the contribution of the ions of the lattice to the polarizability. For instance, an ionic crystal with two atoms for unit cell, with equal mass and charge, that oscillate in an optic transverse normal mode (TO), has an electric dipole momentum, oscillating at the TO frequency. The related dielectric function is

$$\tilde{\varepsilon}(\omega) = \varepsilon_{\infty} + \sum_{k=1}^{N} \frac{4\pi N_k q^2}{M} \frac{1}{\omega_{0k}^2 - \omega^2 - i\gamma_k\omega} \tag{2.60}$$

where M and q are the mass and the charge of the ions, respectively, and ω_{0k} are the vibration frequency of the lattice (associated with the phononic excitations), two or three order of magnitude lower than the electronic transition ones.

Than, the total dielectric function is

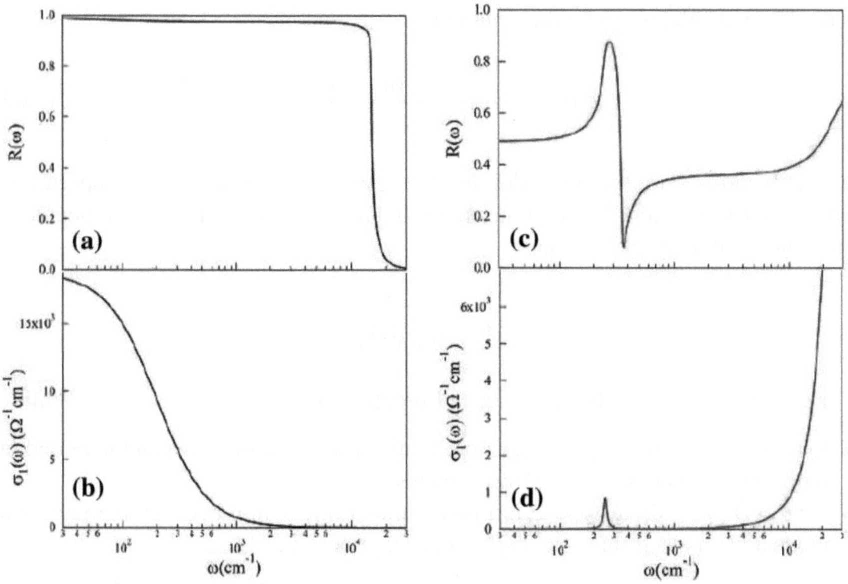

Fig. 2.24 Reflectivity and optical conductivity calculated with the Drude-Lorentz model for: a metal with ω_p = 15,000 cm^{-1}, Γ= 200 cm^{-1} (**a**), (**b**); an insulator with a phonon (I_1^2 = 1,000 cm^{-2}, γ_1 = 20 cm^{-1} and ω_i = 250 cm^{-1}) and an optical electron transition in the visible range (I_2^2 = 100,000 cm^{-2} and ω_2 = 25,000 cm^{-1}) (**c**), (**d**)

$$\tilde{\varepsilon}(\omega) = \varepsilon_\infty - \frac{\omega_p^2}{\omega^2 + i\Gamma_D\omega} + \sum_{j=1}^{N} \frac{S_{el,j}^2}{\omega_{0j}^2 - \omega^2 - i\gamma_j\omega} + \sum_{k=1}^{N'} \frac{S_{ph,k}^2}{\omega_{0k}^2 - \omega^2 - i\gamma_k\omega} \quad (2.61)$$

where $S_{el,j}^2 = \frac{4\pi N_j e^2}{m}$, $S_{ph,k}^2 = \frac{4\pi N_k q^2}{M}$.

In Fig. 2.24a and b the reflectivity and the optical conductivity of a metal are shown as calculated through the Drude model for a good metal with $\omega_p = 15,000\,\text{cm}^{-1}$ and $\Gamma = 200\,\text{cm}^{-1}$. The reflectivity presents a drop at the so called plasma edge frequency and the conductivity shows the Drude peak at $\omega = 0$ having width proportional to Γ. In Fig. 2.24c and d the results for an insulator are shown: one can see a phononic peak ($I_1^2 = 1,000\,\text{cm}^{-2}$, $\gamma_1 = 20\,\text{cm}^{-1}$ and $\omega_i = 250\,\text{cm}^{-1}$) and an optical electron transition in the visible range ($I_2^2 = 100,000\,\text{cm}^{-2}$ and $\omega_2 = 25,000\,\text{cm}^{-1}$).

2.6.2 Fano Model for the Optical Conductivity

The Fano theory describes the interaction of one or more discrete levels with a continuum of states, resulting in asymmetric optical absorption peaks [30]. Therefore, the optical conductivity, after the subtraction of the continuous electronic background, has a Fano profile, given by the equation:

$$\tilde{\sigma}(\omega) = i\sigma_0 \frac{(q-i)^2}{i + (\omega^2 - \omega_T^2)/\gamma\omega} \quad (2.62)$$

where γ and ω_T are the linewidth and the resonant frequency of the unperturbed vibrational state and q is the dimensionless Fano parameter. If θ is the degree of asymmetry of the peak, one has

$$q = \frac{-1}{tg(\theta/2)} \quad (2.63)$$

For $\theta = 0$, or equivalently, $|q| \to \infty$, a Lorentzian line shape is recovered. While

- for $\theta < 0$ ($q > 0$), $\omega_{ph} > \omega_{el}$: a predominant interaction between the phononic mode and an electronic state lower in energy occurs.
- for $\theta > 0$ ($q < 0$), $\omega_{ph} < \omega_{el}$: a predominant interaction between the phononic mode and an electronic state higher in energy occurs.

Here, ω_{ph} is the phonon frequency at $k = 0$ and ω_{el} is the central frequency of the electronic continuous state [31].

The oscillator strength S is also related to the Fano parameter by the relation:

$$S = 4\pi\sigma_0 \frac{(q^2-1)\gamma}{\omega_T^2} \quad (2.64)$$

One can account for the strong oscillator strength observed for the infrared-active phonons, by considering a linear coupling between lattice vibrations and electronic oscillators. For a phonon coupled to an electronic background, the optical conductivity is

$$\sigma_j(\omega) = \frac{-iA\omega}{(1-\lambda_j)\omega_j^2 - \omega(\omega + i\gamma_j)} \quad (2.65)$$

where $A = e_{T,j}^{*2} n_e/\mu$ and $\lambda_j = g^2\epsilon_e/m_e\mu\omega_j^2\omega_{p,e}^2$ is a dimensionless electron-phonon coupling parameter. Moreover,

$$\epsilon_e = \frac{\omega_{p,e}^2}{\omega_e^2 - \omega(\omega + i\gamma_e)}$$

is the contribution to the dielectric function due to an electronic oscillator coupled to the phonons. The indexes e and j refer to the electronic oscillator and the jth phonon, respectively. In addition,

$$e_{T,j}^{*2} = \frac{\lambda_j\epsilon_e n_i (Z_i e)^2 \omega_j 2}{n_e \Omega_{ph}^2}$$

is the transverse effective charge where $\Omega^2_{ph} = 4\pi n_i(Z_i e)^2/\mu$ is the square plasma frequency of the lattice, with Z_i the formal valence of the ions. The Fano asymmetry parameter θ is

$$\theta_j = 2Arg[\epsilon_e(\omega_j)]. \quad (2.66)$$

2.6.3 Fano Resonance in the Extinction Coefficient

In the previous section, we have seen that the Fano theory describes the interference between one or more discrete states (DS) with a continuum of states (CS) (usually an electronic background). In particular, one can observe a Fano resonance when vibrational excitations of a molecule are coupled with a plasmon resonance in surface-enhanced IR absorption [32].

A peculiar characteristic of these resonances is their asymmetric line profile, due to the coexistence of constructive and destructive interference processes. Recently, it has been shown that these phenomena are observable in plasmonic nanostructures [33–37]. In this case the plasmon resonance (PR) acts as the CS and interferes with a DS: the DS can be the excitation of a guided mode, the excitation of a diffraction channel (like a grating or an array), the excitation of a dark (not excited) plasmon mode, the vibrational (active or not IR active) excitation.

Although Fano interference has been known for more than fifty years, there is still a gap between theory and experimental results when plasmonic resonances are involved. Generally, the approach to analyze such a asymmetry is to fit it with a Fano profile (see Sect. 2.6.2), by applying the scattering matrix method and considering all the involved channels as discrete levels. Another possibility consists in using a classical phenomenological method of coupled oscillators.

There is a further method that is able to explain how the energy width of the PR and the energy separation between the PR and the DS, together with their coupling strength, may affect this asymmetric profile. This method can be use for fitting macroscopic optical functions like transmittance and reflectance and will be utilized in Sect. 3.4.2. In Fig. 2.25b one can see a sketch where a plasmonic resonance, with eigenstate $|c\rangle$ and acting as the CS, interferes with a discrete state $|d\rangle$ with a coupling constant v and each one interacts with an incident state (of an electromagnetic field) $|i\rangle$ by g and w, respectively. The result is a new mixed state that accounts for both excitations paths. Performing exact calculations of the probability of exciting that mixed state, one can obtain a simple analytical description of Fano resonances mediated by PRs [32]. Considering the extinction coefficient, defined as

$$\epsilon(\omega) = 1 - T(\omega) \quad (2.67)$$

where $T(\omega)$ is the transmittance, one can write for the shape of the resonance, caused by the coupling between a DS and a CS, the following equation:

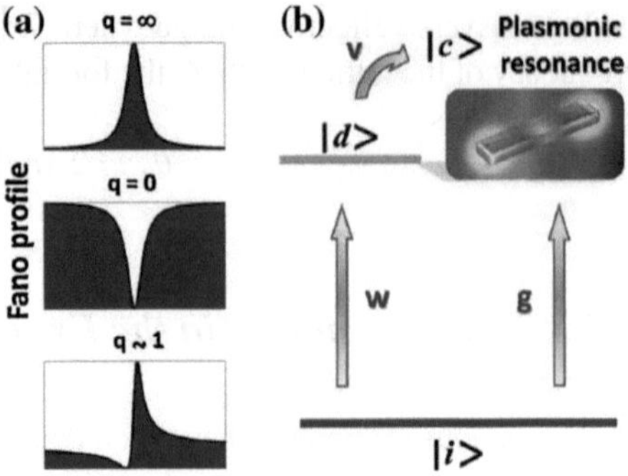

Fig. 2.25 Different Fano line-shapes for three values of the asymmetry parameter q (**a**). Fano process with a plasmonic continuum state: an incident state $|i\rangle$ excites a quasi-continuum state obtained from the interaction of a plasmonic resonance, $|c\rangle$, with a discrete state $|d\rangle$. The interaction is described by the coupling factors w, g and v (**b**) [32]

$$\epsilon(\mathcal{E}) = \frac{(\mathcal{E}+q)^2}{\mathcal{E}^2+1} \tag{2.68}$$

where q is the Fano factor, related also to the excitation probability ratio between the discrete and the continuum state and $\mathcal{E}$ is the reduced energy, defined by

$$\mathcal{E} = \frac{2(\hbar\omega - \hbar\omega_{ph})}{\Gamma_{ph}} \tag{2.69}$$

It depends on

- ω: frequency of the incident photon.
- ω_{ph}: frequency of the discrete state (a phonon in the specific case in Fig. 2.14).
- Γ_{ph}: width of the discrete state.

In Fig. 2.25a the three special case for the Fano profile are reported: $q \rightarrow \infty$, $q = 0$, q finite. In the first case, the probability of directly exciting the continuum is small and the profile is mainly determined by the transition through the discrete mode (Lorentzian shape of, for example, a phonon); in the second case, a symmetric antiresonance arises, known as Breit-Wigner dip [38]; in the third case, an asymmetric line-shape does appear.

In order to include the dependence from the PR feature in Eq. 2.68, now we consider that the coupling to the CS is governed by the excitation of a PR. If $\mathcal{H}_0$ is the unperturbed Hamiltonian, it has an eigenstate $|d\rangle$ with eigenvalue E_{ph} ($\mathrm{E}_{ph} = \hbar\omega_{ph}$) and one, $|c\rangle$, with a continuum spectra of eigenvalues E ($\mathrm{E} = \hbar\omega$). If, then, V is the Hamiltonian which couples $|d\rangle$ with $|c\rangle$, one can write (supposing $\hbar = 1$):

$$\langle d|\mathcal{H}_0|d\rangle = \omega_{ph} \tag{2.70}$$
$$\langle c|\mathcal{H}_0|c'\rangle = \omega\delta(\omega - \omega') \tag{2.71}$$
$$\langle c|V|d\rangle = v\sqrt{\mathcal{L}(\omega)} \tag{2.72}$$
$$\langle d|V|d\rangle = 0 \tag{2.73}$$
$$\langle c|V|c\rangle = 0 \tag{2.74}$$

One can see how the coupling between $|c\rangle$ and $|d\rangle$ is determined both by the coupling constant v and by the plasmonic line-shape $\mathcal{L}(\omega)$, which is supposed to be Lorentzian, that is

$$\mathcal{L}(\omega) = \frac{1}{1 + \left(\frac{\omega-\omega_p}{\Gamma_p/2}\right)^2} \tag{2.75}$$

where

- ω_p is the frequency of the plasmon resonance.
- Γ_p is the width of the plasmon resonance.

If $\mathcal{H} = \mathcal{H}_0 + V$ is the total Hamiltonian, we can solve the eigenvalue problem $\mathcal{H}|\Psi\rangle = E|\Psi\rangle$, where $|\Psi\rangle$ is the new mixed state quasi-CS. If, then, we consider an incident photon in the DS $|i\rangle$, we have a coupling by an Hamiltonian W with the state $|d\rangle$ and $|c\rangle$, that is

$$\langle i|W|d\rangle = w \tag{2.76}$$

$$\langle i|W|c\rangle = g\sqrt{\mathcal{L}(\omega)} \tag{2.77}$$

where w and g are the coupling factors.

We can recover the Fano profile by solving the previous problem, i.e. by calculating the probability $|\langle i|W|\Psi\rangle|^2$ that a photon in state $|i\rangle$ excites a quasi-CS state $|\Psi\rangle$. Hence, by normalizing the latter result with the probability of exciting the continuum PR in the absence of the DS, we obtain the same result as in Eq. 2.68, namely

$$\frac{|\langle i|W|\Psi\rangle|^2}{|\langle i|W|c\rangle|^2} = \frac{(\mathcal{E}+q)^2}{\mathcal{E}^2+1} \tag{2.78}$$

with, now, q and $\mathcal{E}$ related also to the width and the energy position of the PR, that is

$$q(\mathcal{E}) = \frac{vw/g}{\Gamma_{ph}(\omega)/2} + \frac{\omega-\omega_p}{\Gamma_p/2} \tag{2.79}$$

$$\mathcal{E} = \frac{\omega}{\Gamma_{ph}(\omega)/2} - \frac{\omega-\omega_p}{\Gamma_p/2} \tag{2.80}$$

Here, in particular,

$$\Gamma_{ph}(\omega) = 2\pi v^2 \mathcal{L}(\omega) \tag{2.81}$$

relating the energy width of the DS to the coupling constant v to the PR line shape. In particular this means that the lifetime $1/\Gamma_{ph}$ of the discrete state is completely determined by its coupling with the continuous state. When V is small compared with $\mathcal{H}_0$, $\Gamma_{ph}(\omega)$ coincides with the decay rate of the DS: in absence of interaction between DS and CS, if DS is an IR active mode, it should have a not zero width Γ'_{ph}.

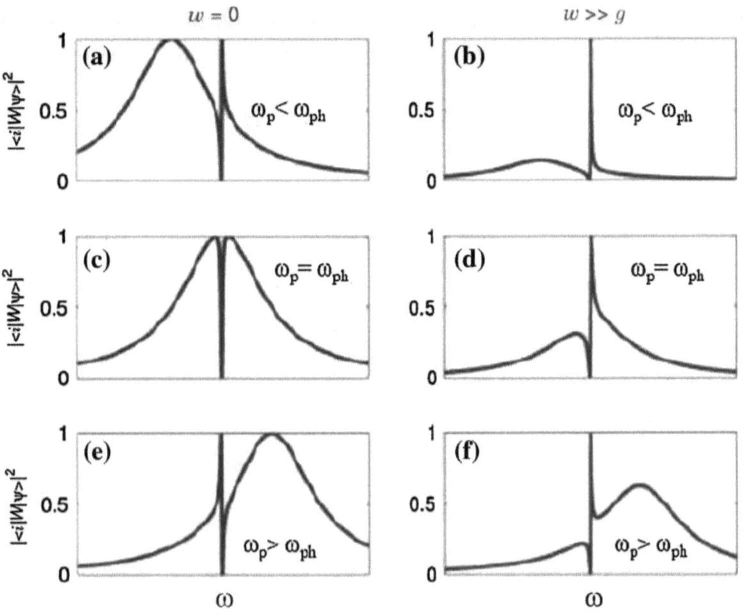

Fig. 2.26 Fano resonances calculated from Eq. 2.78–2.79 for two different coupling regimes $w = 0$ (**a**), (**c**), (**e**) and for $w \gg g$ (**b**), (**d**), (**f**), when a DS with frequency ω_{ph} interacts with a continuum plasmon state with frequency ω_p ($\Gamma_p = 10\Gamma_{ph}$) for different relative positions [32]

The dependence of q and $\mathcal{E}$ on the parameter $\omega, \omega_p, \Gamma_p, \Gamma_{ph}$ and the three coupling factors, v, w e g, explains why the Fano resonance exhibits different degrees of asymmetry.
One can distinguish two main cases (see Fig. 2.26):

- $w \ll v, g$: the line-profile is mainly determined by the PR, which is modified by the coupling to the DS (v). Indeed, the DS mainly excites indirectly through the plasmonic state. As the minimum of the Fano resonance always lies between the two maxima, the one of the DS (phononic peak) and the other one of the plasmonic state, in the case of $w = 0$ and $\omega_p = \omega_{ph}$ a symmetric dip is obtained.
- $w \gg g$: the interaction with the CS mainly goes through the DS, then the main obtained resonance is that of the DS (phononic peak), particularly in the case $\omega_p < \omega_{ph}$.

Therefore, it is rather simple to understand which coupling regime holds. This model will be used in Chap. 3 to explain the plasmonic spectra observed in topological patterned films of Bi_2Se_3.

References

1. J.S. Nodvick, D.S. Saxon, Phys. Rev. **96**, 180 (1954)
2. R.L. Warnock, P. Morton, Particle Accelerators **25**, 113 (1990)
3. J.B. Murphy, S. Krinsky, R.L. Gluckstern, Particle Accelerators, **57**, 9 (1997)
4. M. Abo-Bakr, J. Feikes, K. Holldack, G. Wüstefeld, H.-W. Hübers, Phys. Rev. Lett. **90**, 254801 (2003)
5. M. Abo-Bakr, J. Feikes, K. Holldack, P. Kuske, W.B. Peatman, U. Schade, G.Wüstefeld, H.-W. Hübers. Phys. Rev. Lett. **90**, 094801 (2003)
6. M. Venturini, R. Warnock, Phys. Rev. Lett. **89**, 224802 (2002)
7. A.J. LaRocca, in *The Infrared Handbook*, ed. by G. Zissis, W. Wolfe for IRIA, chap2 (1978)
8. F.J. Studer, R.F. Van Beers, J. Opt. Soc. Am. **54**, 945 (1964)
9. J.E. Chamberlain, G.W. Chantry, F.D. Findlay, H.A. Gebbie, J.E. Gibbs, N.W.B. Stone, A.J. Wright, Infrared Phys. **6**, 195 (1966)
10. C.C. Homes, M. Reedik, D.A. Cradles, T. Timusk, Appl. Optics **32**, 2976 (1993)
11. F. Wooten, *Optical Properties of Solids* (Academic Press, New York, 1972)
12. M. Dressel, G. Grüner, *Electrodynamics of Solids* (Cambridge University Press, Cambridge, 2002)
13. H.A. Kramers, Nature (London) **117**, 775 (1926)
14. H.A. Kramers, in *Estratto dagli Atti del Congresso Internazionale di Fisica*, vol. 2 (Zanichelli, Bologna, 1927), p. 545
15. H.A. Kramers, *Collected Scientific Papers* (North-Holland, Amsterdam, 1956)
16. R. de L. Kronig, J. Opt. Soc. Am. **12**, 547 (1926)
17. R. de L. Kronig, Ned. Tjidschr. Natuurk. **9**, 402 (1942)
18. P.F. Henning, C.C. Homes, S. Maslov, G.L. Carr, D.N. Basov, B. Nikolić, M. Strongin, Phys. Rev. Lett. **83**, 4880 (1999)
19. S.A. Maier, *Plasmonics: Fundamentals and Applications* (Springer, New York, 2007)
20. P.W. Milonni, *Fast Light, Slow Light and Left-Handed Light* (Institute of Physics Publishing, Bristol, 2005)
21. W.L. Barnes, A. Dereux, T.W. Ebbesen, Nature **424**, 824 (2003)
22. W. Zhang, Eur. Phys. J. Appl. Phys. **43**, 1 (2008)
23. G. Goubau, J. Appl. Phys. **21**, 1119 (1950)
24. J.B. Pendry, L. Martin-Moreno, F.J. Garcia-Vidal, Science **305**, 847 (2004)
25. Y.S. Hor A. Richardella, P. Roushan, Y. Xia, J.G. Checkelsky, A. Yazdani, M.Z. Hasan, N.P. Ong, R.J. Cava, Phys. Rev. B **79**, 195208 (2009)
26. S. Jia, H. Ji, E. Climent-Pascual, M.K. Fuccillo, M.E. Charles, J. Xiong, N.P. Ong, R.J. Cava, Phys. Rev. B **84**, 235206 (2011)
27. R. Valdés Aguilar, A.V. Stier, W. Liu, L.S. Bilbro, D.K. George, N. Bansal, L. Wu, J. Cerne, A.G. Markelz, S. Oh, N.P. Armitage, Supplemental Materials for: Phys. Rev. Lett. **108**, 087403 (2012)
28. N. Bansal, Y.S. Kim, M. Brahlek, E. Edrey, S. Oh, Phys. Rev. Lett. **109**, 116804 (2012)
29. N.W. Ashcroft, D. Mermin, *Solid State Physics* (Saunders College Publishing, New York, 1976)
30. U. Fano, Phys. Rev. **124**, 1866 (1961)
31. A. Damascelli, K. Schulte, D. van der Marel, A.A. Menovsky, Phys. Rev. B **55**, R4863 (1997)
32. V. Giannini, Y. Francescato, H. Amrania, C.C. Phillips, S.A. Maier, Nano Lett. **11**, 2835 (2011)
33. F. Hao, Y. Sonnefraud, P. Van Dorpe, S.A. Maier, N.J. Halas, P. Nordlander, Nano Lett. **8**, 3983 (2008)
34. N. Verellen, Y. Sonnefraud, H. Sobhani, F. Hao, V.V. Moshchalkov, P. Van Dorpe, P. Nordlander, S.A. Maier, Nano Lett. **9**, 1663 (2009)
35. L.A. Fan, C. Wu, K. Bao, J. Bao, R. Bardhan, N.J. Halas, V.N. Manoharan, P. Nordlander, G. Shvets, R. Capasso, Science **328**, 1135 (2010)
36. I.M. Pryce, K. Aydin, Y.A. Kelaita, R.M. Briggs, H.A. Atwater, Nano Lett. **10**, 4222 (2010)
37. Y. Sonnefraud, N. Verellen, H. Sobhani, G. Vandenbosch, V. Moshchalkov, P. Van Dorpe, P. Nordlander, S.A. Maier, ACS Nano **4**, 1664 (2010)
38. A. Miroshnichenko, S. Flach, Y. Kivshar, Rev. Mod. Phys. **82**, 2257 (2010)

Chapter 3
Results and Analysis

Abstract In this chapter the experimental results of this thesis will be shown and the related analysis explained. Four topological insulator crystals (Bi_2Se_3, $Bi_{2-x}Ca_xSe_3$, Bi_2Se_2Te and Bi_2Te_2Se) have been optically studied by FTIR Spectroscopy, with increasing chemical compensation. They have been measured from 5 to 300 K and from subterahertz to visible frequencies. The effect of compensation is clearly observed in the infrared spectra through the suppression of the extrinsic Drude term together with the appearance of strong absorption peaks, that we assign to electronic transitions among localized impurities states. From the far-infrared spectral weight of the most compensated sample (Bi_2Te_2Se), one can estimate a density of charge carriers on the order of 10^{17} cm10^{-3} in good agreement with transport data. Those results demonstrate that the low electrodynamics in single crystals of TI, even at the highest degree of compensation presently achieved, is still influenced by three-dimensional charge excitations. Its spectral weight is, indeed, still nearly higher by two orders of magnitude than that expected from the topological surface states, estimated from optical conductivity of films of Bi_2Se_3 on sapphire substrate. Such films have been measured in the sub-THz and THz frequency region, in order to study their optical conductivity as a function of their thickness. One can observe no appreciable change in the free carriers contribution, while the α phonon intensity strongly decreases with decreasing thickness, demonstrating that the only contribution to the transport is due to surface carriers, not depending on bulk characteristics.The surface metallic state of the thin TI films has been finally studied by patterning the films by a grating, as explained in Chap. 2. This provides the possibility to detect surface plasmonic collective modes, due to the excitation of two dimensional charge density waves along the topological interface of the samples. In the last part of this thesis those plasmons will be analyzed, demonstrating that they have two dimensional nature characteristic of 2DEGs (see Sect. 1.2.1).

P. Di Pietro, *Optical Properties of Bismuth-Based Topological Insulators*,
Springer Theses, DOI: 10.1007/978-3-319-01991-8_3,

3.1 Spectra of the Crystalline Topological Insulators

The investigation of charge transport and cyclotron resonances of Dirac quasi-particles in as-grown Bi_2Se_3, Ca doped Bi_2Se_3 ($Bi_{2-x}Ca_xSe_3$) and alloys Bi_2Se_2Te and Bi_2Te_2Se, also by IR spectroscopy, has proven to be challenging because the surface current contribution is usually obscured by extrinsic bulk carriers response [1–3].

Indeed, as-grown crystals of Bi_2Se_3 display a finite density of Se vacancies (see Sect. 1.3.1), which act as electron donors. They pin the bulk chemical potential within the conduction band, thus producing over a wide range of carrier concentrations, extrinsic n type degenerate semiconductor behavior. Therefore, Se vacancies also affect the low-energy transport properties of those materials [4], making it difficult to distinguish the intrinsic metallic behavior due to the topological surface state from the extrinsic metallic conduction induced by the Se non-stoichiometry . As a consequence, both transport and optical conductivity experiments [2, 5] show a metallic behavior with a Drude term confined at low frequencies ($\omega < 600$ cm^{-1}) which mirrors the extrinsic carrier content.

Two phonon peaks, one of which clearly shows an asymmetric Fano shape (the α-mode) which indicates an interaction with the electronic continuum, have been observed in the far-infrared (FIR) range (see Sect. 1.3.4) [2, 6]. The bulk insulating gap spans between 250 and 350 meV, depending on the Se vacancy content, in good agreement with theoretical calculations [7] (see Sect. 1.3.1) . Motivated by the transport characteristics described in Sect. 1.3.2, materials with a reduced non-stoichiometry induced by bulk carriers, have been grown and then chosen for the present optical investigation. It has been showed (Sects. 1.3.1 and 1.3.2) for instance that Ca doping in the Bi site ($Bi_{2-x}Ca_xSe_3$) progressively shifts the chemical potential from the conduction band to the valence band, making the material a p type degenerate semiconductor. Furthermore, by exploiting the different doping chemistry of Bi_2Se_3 (n type) and Bi_2Te_3 (p type) a better compensation was obtained in the Bi_2Se_2Te and Bi_2Se_2Te alloys [4, 8, 9]. In the latter, a high resistivity at low temperature (exceeding 1 Ωcm) was observed. Such that (ab)-plane resistivity shows an increasing (semiconducting) behavior down to about 50 K followed by a low-T regime, in which resistivity saturates at values exceeding 1 Ωcm: in this regime, surface charge carrier mobility, much higher than the bulk mobility, has been experimentally detected [9, 10] (see Sect. 1.3.2 for further details). Moreover, in Bi_2Se_2Te the variable range-hopping behavior expected for an impurity-driven conductivity was found to give place, below 20 K, to a T-independent dc-conductivity. This crossover was reported as providing evidence that surface conductance prevails at low T in the best compensated system [8].

However, the effects of chemical compensation on the optical properties of those materials have not been investigated up to now. In this section the optical properties of the four TI crystals Bi_2Se_3, $Bi_{2-x}Ca_xSe_3$, with x = 0.0002, Bi_2Se_2Te and Bi_2Te_2Se will be presented. The effects of the enhanced compensation will be clearly visible in the FIR spectra through the suppression of the Drude term and the appearance of strong absorption peaks, that we assign to electronic transitions among localized states, similar to those found in weakly doped semiconductors. Our data show that the electrodynamics at low energy of the most compensated sample Bi_2Te_2Se is still affected by 3D doped charges, as therein the FIR spectral weight is higher by nearly two orders of magnitude than the spectral weight associated with topological states of a thin film of Bi_2Se_3 on sapphire substrate (see Sect. 3.3).

3.1.1 Reflectivity

The reflectivity $R(\omega)$ of the four single crystals was measured at near-normal incidence with respect to the *ab* basal plane from sub-THz to visible frequencies (*i.e* from about 10 to 22500 cm^{-1} or 1 meV $\div$ 2.8 eV) at temperatures ranging from 5 to 300 K, shortly after cleaving the sample, in order to achieve the lowest possible roughness of its surface.

The dimensions of the *ab* surfaces were 2×10 mm^2 for Bi_2Se_3, 2×3 mm^2 for Bi_2Se_2Te and $Bi_{1.9998}Ca_{0.0002}Se_3$ and 5×5 for Bi_2Te_2Se, while their thickness along the c axis was about 1 mm, 1 μm, 1 μm and 4 mm, respectively. The measurements were performed by means of the experimental setup described in Sect. 2.2 and by using below 30 cm^{-1} the coherent synchrotron radiation of BESSY II (see Sect. 2.1.1).

The reflectivity data are reported in Fig. 3.1 for the four crystals at all temperatures. In all spectra a strong absorption appears above 10000 cm^{-1}. In Fig. 3.1a, b all crystals display a plasma edge around 500 cm^{-1}, which confirms the picture of extrinsic transport in these materials. In $Bi_{1.9998}Ca_{0.0002}Se_3$ (Fig. 3.1c), Ca doping shifts the plasma edge to about 400 cm^{-1}. Finally, in Bi_2Te_2Se (Fig. 3.1d) the plasma edge further shifts to about 200 cm^{-1} indicating that the strongest compensation is achieved. Moreover in the latter sample, unshielded phonons are well resolved at about 60 (α mode) and 130 (β mode) cm^{-1} which will be discussed in the next Section. The reflectivity is almost independent of temperature, except for a slight softening of the plasma edge (and a narrowing of the phonon absorption in panel (d)) as $T \rightarrow 0$.

3.1.2 Optical Conductivity

The optical conductivity $\sigma_1(\omega)$ obtained from the $R(\omega)$ in Fig. 3.1 by Kramers-Kronig transformations is shown for the same temperatures and frequencies in Fig. 3.3. The direct-gap transition, which corresponds to a small bump around 3000 cm^{-1}, is barely visible due to its superposition to the huge electronic excitation above 10000 cm^{-1} (about 1.2 eV). This peak actually has a triplet structure, in which one can distinguish three peaks, labeled with E_1, E_2 and E_3. In particular reflectivity measurements reported in Ref. [12] showed that E_2 has a double fine

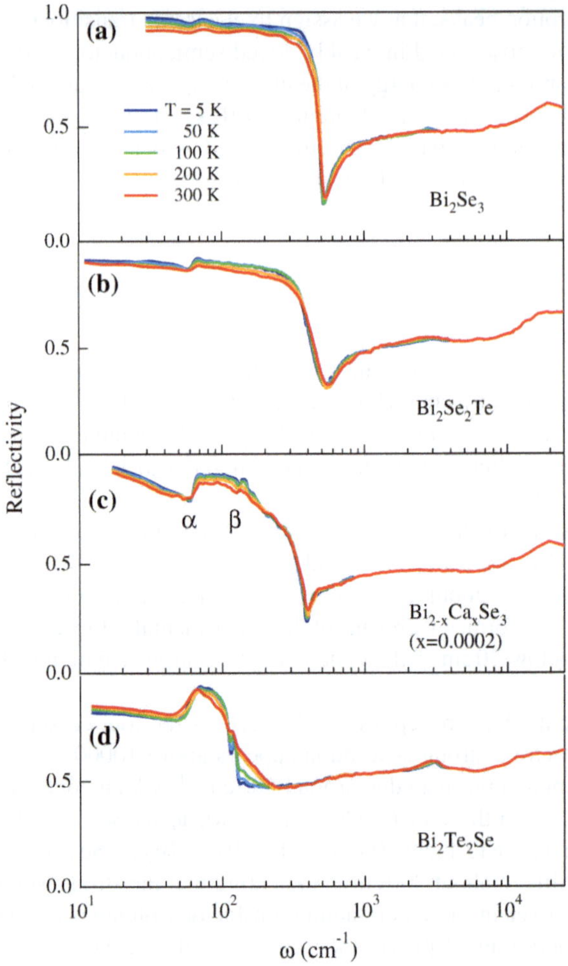

Fig. 3.1 Reflectivity of Bi_2Se_3 (**a**), Bi_2Se_2Te (**b**), $Bi_{1.9998}Ca_{0.0002}Se_3$ (**c**) and Bi_2Te_2Se (**d**) from 10 to 24000 cm^{-1} at different temperatures. The FIR spectra in (**a**)(**c**) are characterized by a free-carrier plasma edge around 500 cm^{-1} (400 ,cm^{-1}) as well as phonon features α and β at about 60 and 130 cm^{-1}, respectively. In (**d**), due to a strong compensation, the phonon absorption is best observed. In all spectra, a weak bump develops at low T around 3000 cm^{-1}, corresponding to the direct-gap transition. The triplet direct gap appears instead above 10000 cm^{-1}. The α- and β-infrared-active phonon modes are indicating in panel (**c**) [11]

structure mostly observable at low temperature. The energy of these three peaks are 2.24, 3.73 (and 4.24 eV) and 4.87 eV, respectively. They correspond to an optical gap E_0, which arises from transitions between bands having different symmetry.

Most of the effects induced by compensation appear below the plasma edge at 500 cm^{-1}. The most extrinsic system, Bi_2Se_3 (Fig. 3.3a) presents a Drude term superimposed to the α- and β-phonon peaks, which both sharpen for decreasing T.

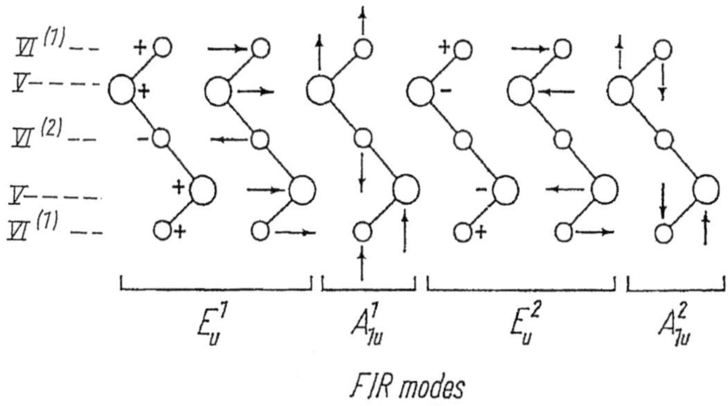

Fig. 3.2 Infrared-active modes of the rhombohedral V_2 (Bi) VI_3 (Se) compounds. The superscripts 1, 2 of the E_u (|| to the *ab* plane) and A_{1u} (|| to the *c* axis) representations correspond to the low- and high- frequency modes, respectively, that is E_u^1 is the α-mode and E_u^2 is the β-mode [6]

A similar behavior has been observed in Ref. [2] on a crystal with a comparable charge-carrier density ($\sim 10^{18}$ cm^{-3}, as described in Sect. 1.3.4 and resumed below.) The effect of compensation becomes observable in Bi_2Se_2Te (Fig. 3.3b). Here, at variance with an appreciable dc conductivity ($\sigma_{dc} \sim 200\Omega^{-1}$cm^{-1}), most of the FIR spectral weight is located at finite frequencies in the phonon spectral region.

A further drastic reduction in the spectral weight is finally obtained in $Bi_{1.9998}Ca_{0.0002}Se_3$ and Bi_2Te_2Se (Fig. 3.3c, d), where the Drude term is strongly suppressed.

The rhombohedral structure of the undoped Bi_2Se_3 is centro-symmetric. Group theory predicts ten zone-center modes corresponding to the following irreducible representations [13]:

$$10 = 2A_{1g} + 3A_{1u} + 2E_g + 3E_u$$

The three acoustic branches come from one A_{1u} and a doubly degenerate E_u, while the rest corresponds to optical modes. The *gerade* (g) modes are Raman active, while the *ungerade* (u) ones are IR active. Therefore, there are four Raman-active modes ($2A_{1g}$+$2E_g$) and four IR-active modes ($2A_{1u}$+$2E_u$). As shown in Fig. 3.2, the E_u phonons correspond to atomic vibrations in the plane of the layers , while the A_{1u} modes correspond to vibrations along the *c* axis perpendicular to the layers [13, 14].

Being the α mode at lower frequency with respect to the β mode, it is related to the mode labelled with E_u^1, while the other is related to E_u^2.

The phononic contribution to the optical conductivity can be separated to better characterized the extrinsic one, that we will describe in the next section.

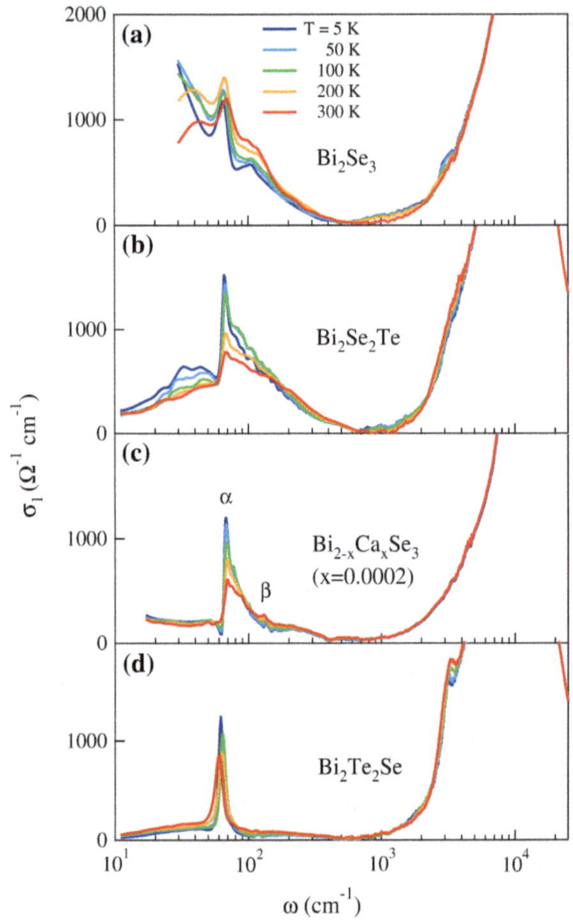

Fig. 3.3 Optical conductivity of Bi_2Se_3 (**a**), Bi_2Se_2Te (**b**), $Bi_{1.9998}Ca_{0.0002}Se_3$ (**c**) and Bi_2Te_2Se (**d**) from 10 to 24000 cm^{-1} at different temperatures. A broad minimum around 500 cm^{-1} separates the high-frequency interband transitions from the low-energy excitations. The α- and β-infrared-active phonon modes are indicated in panel (**c**) [11]

3.1.3 Extrinsic Contributions to the Optical Conductivity

In the insets of Fig. 3.4 examples of those fits (dotted lines) are shown at 5 K (see for more details Sect. 3.1.4 below).

The $\sigma_1(\omega)$ in the FIR region is shown in the main panels of Fig. 3.4, as obtained after subtraction of both the interband and the phonon contributions by a Drude-Lorentz (D-L) fit (see Sect. 2.6.1).

The electronic conductivity of the undoped sample Bi_2Se_3 (Fig. 3.4a) can be described in terms of a Drude term (open circles), which narrows for decreasing T in

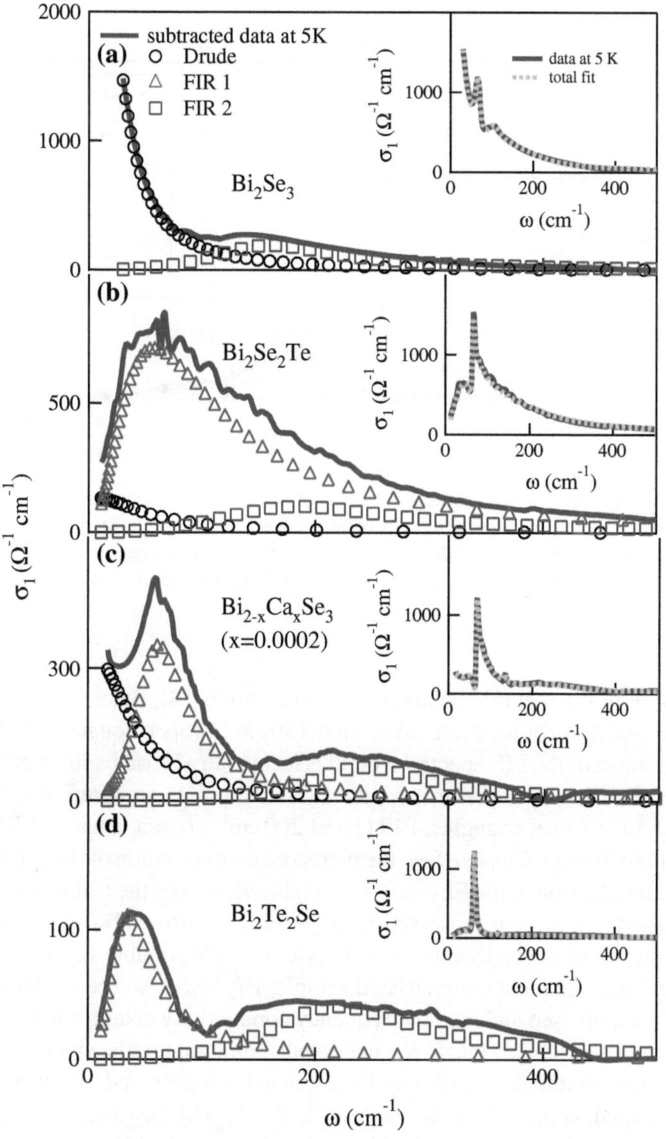

Fig. 3.4 FIR optical conductivity of Bi_2Se_3 (**a**), Bi_2Se_2Te (**b**), $Bi_{1.9998}Ca_{0.0002}Se_3$ (**c**) and Bi_2Te_2Se (**d**) at 5 K, after subtraction of both the interband and the phonon contributions via Drude-Lorentz fits. The Drude term and the FIR contributions 1 and 2 are indicated by *open circles*, *triangles* and *squares*, respectively. Note the different vertical scales in each panel. In the insets the fit (*dotted lines*) and data at 5 K are shown [11]

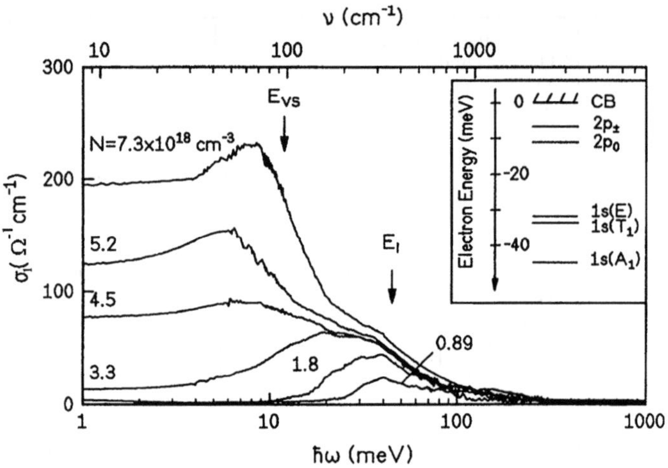

Fig. 3.5 Optical conductivity of different samples of Si:P (different carrier concentration) at 10 K. The inset gives a sketch of the level scheme of Si:P states in the dilute limit, including valley-orbit splitting of 1s states; CB is the conductive band, E_I is the ionization energy of the impurity atom and E_{VS} is the valley-orbit splitting between the $1s(A_1)$ and the closely spaced $1s(T_1)$ and $1s(E)$ levels [15]

agreement with the metallic behavior of the resistivity [4], described in Sect. 1.3.2, and of a broad absorption centered around 150 cm^{-1} (open squares). In Bi_2Se_2Te, (Fig. 3.4b), most of the FIR spectral weight is located in a broad band centered around 100 cm^{-1}. This band has been modeled by two Lorentzian contributions, peaked around 50 cm^{-1} (open triangles, FIR1) and 200 cm^{-1} (open squares, FIR2). In the doped sample, $Bi_{1.9998}Ca_{0.0002}Se_3$, the increased compensation of the carriers results in an overall reduction of the FIR spectral weight. Moreover, the FIR absorption splits into two bands, as already observed in Bi_2Se_2Te: a narrow absorption centered at about 50 cm^{-1} and a broader one around 200 cm^{-1}. This double spectral structure is also observed in the most compensated sample, Bi_2Te_2Se, where the Drude term is completely suppressed and the low frequency conductivity takes a value comparable to the $\sigma_{dc} \sim 1\,\Omega^{-1}\,cm^{-1}$, measured in crystals belonging to the same batch [9].

Similar low-frequency absorption bands have been observed on other more conventional doped semiconductors, such as Si:P [15, 16] (see Fig. 3.5). Therein, an insulator-to-metal transition IMT of Mott-Anderson type [17] can be observed for a charge carrier density $n_{IMT} \sim 3.7 \times 10^{18}$ [15–17]. Such transition occurs in an insulating system when the doping concentration increases and consequently the mobility of electrons decreases due to scattering by the ionized impurities. The transitions of isolated P impurities, in Si:P, are the hydrogenlike $1s \rightarrow np$, and at low doping produce narrow peaks in the absorption coefficient. Those peaks are followed by a broad band at higher frequencies due to the transitions from the impurity bound states to the continuum. The narrow peaks broaden for increasing doping, giving rise to a low-frequency band , which remains distinguished from the higher-frequency

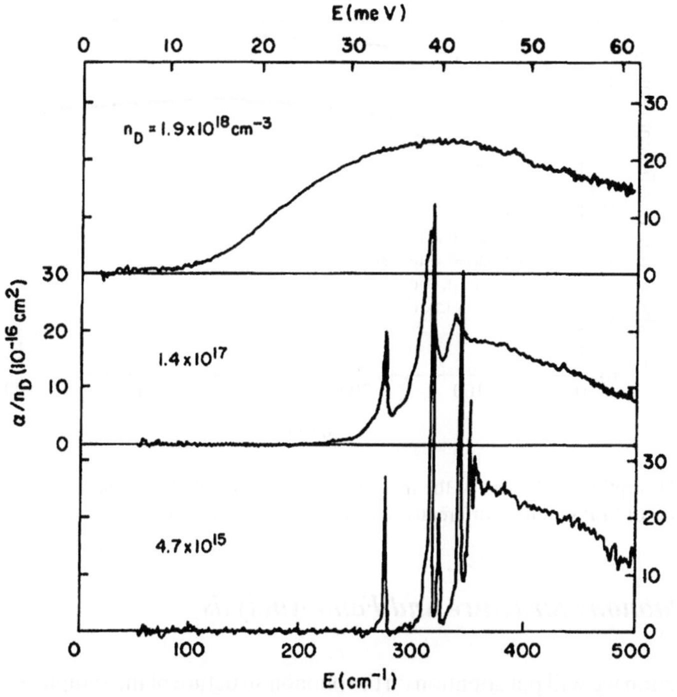

Fig. 3.6 Absorption coefficient (normalized to the donor densities n_D) as a function of frequency for three widely separated donor densities in sample of Si:P at about 2 K [16]

absorption. When the donor density achieves a critical value, such impurity band, formed by the degenerate and unresolved broaden transition peaks, merges with the conduction band, transforming in a Drude term. In this situation the system has become metallic.

Here, in the insulating phase ($n < n_{IMT}$), the FIR2 band at 200 cm^{-1} in $Bi_{1.9998}Ca_{0.0002}Se_3$ and Bi_2Te_2Se, in analogy with Si:P, can be assigned to the transitions from the impurity bound states to the electronic continuum. The FIR2 band is also in very good agreement with the impurity ionization energy estimated from the T dependence of resistivity and Hall data, namely, $E_i \sim$ 20–40 meV [8, 9, 18]. Instead, the low-frequency FIR1 band, clearly resolved in both $Bi_{1.9998}Ca_{0.0002}Se_3$ and Bi_2Te_2Se, as shown in Fig. 3.4c, d, respectively, can be associated with hydrogenlike $1s \rightarrow np$ transitions, broadened by the inhomogeneous environment of the impurities and/or by their interactions.

Furthermore, in Fig. 3.7 we report the absorption coefficient at 5 K of the most compensated sample in comparison with that one of Si:P samples, some of those are also shown in Fig. 3.6.

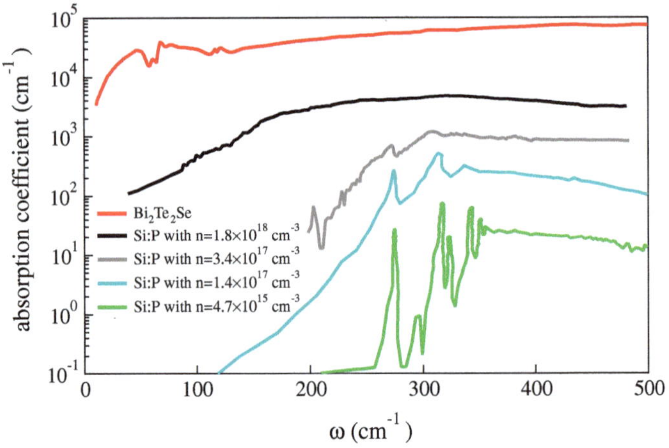

Fig. 3.7 Absorption coefficient of the most compensated sample Bi_2Te_2Se at 5 K in comparison with that one of Si:P samples, differently doped, from Ref. [15] (some *curves* as in Fig. 3.6)

3.1.4 Phonons Structure and Fano Analysis

In this section we will put attention on the phonon structure of the sample, in particular on the Fano character of the α-mode.

Indeed, both in Bi_2Se_2Te and $Bi_{1.9998}Ca_{0.0002}Se_3$, the α-phonon mode shows a Fano line shape (see Sect. 2.6.2) with a low frequency dip, more pronounced at low temperatures. This suggests an interaction of this mode with an electronic continuum at lower frequencies [19]. The Fano shape is much less evident in Bi_2Te_2Se, where the α phonon shows a nearly Lorentzian shape at room temperature and a weak low-frequency dip at low T. This indicates a transfer of the electronic continuum spectral weight from above to below the phonon frequency for increasing temperature (see Sect. 3.1.4 below) [20]. At variance with previous samples, the α mode in Bi_2Se_3 shows, instead, a high frequency dip at all temperatures, in agreement with the observation reported in Ref. [2]. The behavior suggests that the electronic continuum is located in this sample, on average, at higher frequencies, with respect to the α-phonon central frequency.

In Fig. 3.8 we report D-L fits to the optical conductivity of all the four samples—in the far-infrared range at the lowest and the highest temperature—where moreover the α-phonon mode is described in terms of a Fano shape. In Table 3.1 the fitting parameters for the Drude term and the α phonon mode are reported instead for all the measured temperatures.

The most metallic sample, Bi_2Se_3, (Fig. 3.8a) shows a well defined Drude term and a high frequency dip at both temperatures, even if it seems more pronounced at 5 K. Actually, since that feature suggests an interaction between the discrete state (α-mode) and a continuum located at higher frequencies (see Sect. 2.6.2 which may be assigned to the FIR2 band described above, such interaction changes with increasing

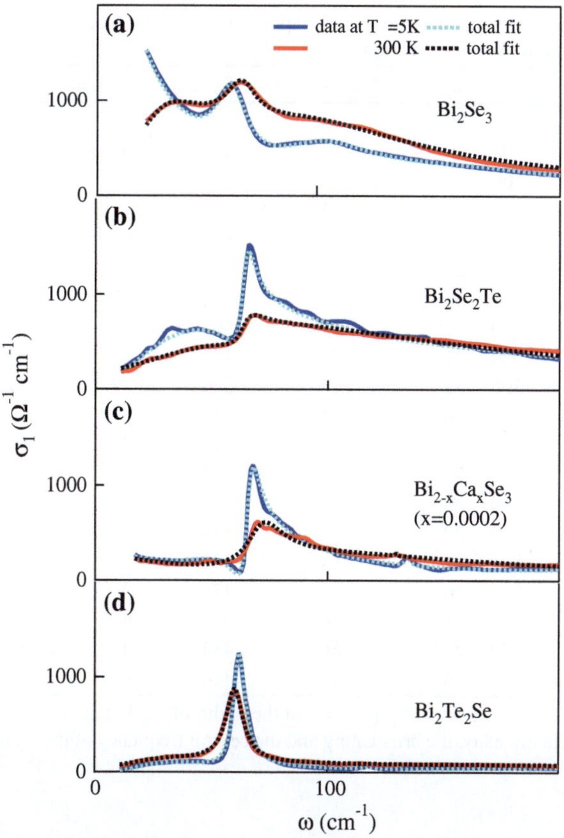

Fig. 3.8 FIR optical conductivity of Bi_2Se_3 (**a**), Bi_2Se_2Te (**b**), $Bi_{1.9998}Ca_{0.0002}Se_3$ (**c**) and Bi_2Te_2Se (**d**) from 0 to $200\,cm^{-1}$ at the lower and higher temperatures measured and the related D-L fits with a Fano shape for the α mode (*dotted line*)

temperature. Indeed, the spectral weight moves from lower to higher frequencies, that is from the Drude term to the FIR1 band, making the interaction somewhat more pronounced: the α phonon interacts with a better defined continuous band and the Fano factor (q) achieves a smaller modulus at room temperature (if $|q| \rightarrow \infty$ the phononic mode recovers a Lorentzian shape).

Both in Bi_2Se_2Te and $Bi_{1.9998}Ca_{0.0002}Se_3$ (Fig. 3.8b, c) the α-phonon mode shows a Fano line shape with a low frequency dip, more pronounced at low temperatures. In those samples, the interaction occurs between the phonon and a continuum located at lower frequency: probably this latter is the FIR1 band (see above) whose spectral weight increases at low temperature. The q factor modulus slightly increases at high temperature, even if in all those three samples it remains quite constant as a function of T (see Fig. 3.9).

Table 3.1 Drude-Lorentz fitting with Fano shape for the α-mode parameters for $\sigma_1(\omega)$ spectrum of the four samples at all the temperatures

	T (K)	$\omega_{p,D}$ (cm^{-1})	γ_D (cm^{-1})	S_α (cm^{-1})	γ_α (cm^{-1})	ω_α (cm^{-1})	q
Bi_2Se_3	5	49	18	506	7	66	-12 ± 1
	50	49	34	441	6	66	-9 ± 1
	100	48	42	451	7	67	-10 ± 1
	200	42	214	465	7	68	-8 ± 2
	300	40	241	398	8	70	-9 ± 2
Bi_2Se_2Te	5	26	57	502	3	64	3 ± 1
	50	40	217	602	5	65	3 ± 1
	100	42	229	561	5	65	3 ± 1
	200	41	241	377	5	64	3 ± 2
	300	39	186	326	5	64	3 ± 2
$Bi_{1.9998}Ca_{0.0002}Se_3$	5	30	41	531	3	66	3 ± 1
	50	28	17	446	3	66	2 ± 1
	100	28	20	483	4	66	2 ± 1
	200	30	38	502	5	69	4 ± 2
	300	29	40	559	9	70	6 ± 2
Bi_2Te_2Se	5	8	14	495	3	62	15 ± 1
	50	16	20	472	3	63	15 ± 1
	100	21	30	520	4	64	15 ± 1
	200	21	60	482	4	63	26 ± 2
	300	18	41	454	4	60	99 ± 2

Here $\omega_{p,D}$ and γ_D are the plasma frequency and the width of the Drude term, respectively; S_α, γ_α and ω_α indicate the intensity, the broadening and the central frequency of the main phononic mode; q is the Fano factor

In Bi_2Te_2Se (Fig. 3.8d) the situation is rather different: the Fano shape is less evident than in the other cases, suggesting that the interaction with the low frequency continuous band is decreased. That band has indeed a smaller spectral weight with respect to the phonon and thus the interaction just provides a weak low-frequency dip at low T and a nearly Lorentzian shape at room temperature. As one can observe in Fig. 3.9, indeed, for Bi_2Te_2Se the q factor is an increasing function of T, with a high modulus.

3.2 Spectra of the Thin Films of Topological Insulators

Given that the topological effects at the surface of single crystals appeared to be masked by residual bulk conductivity, we have focused our interest on the IR spectroscopy of TI thin films, where the ratio of the surface contribution to that of the bulk is more favorable. We have selected two thin films of Bi_2Se_3 on a sapphire substrate (Al_2O_3) with thickness of 60 QL (1 QL $\sim$ 1 nm) and 120 QL. We have measured their

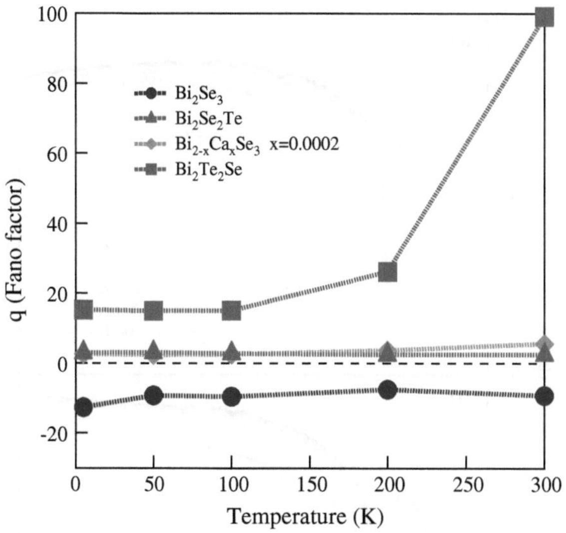

Fig. 3.9 Fano factor (q) as a function of the temperature for the four samples. The error bars are within the symbol size

transmittance in the sub-THz and THz regions i.e from 7 cm^{-1} to about 300 cm^{-1} [1] for different temperatures. As described in Sect. 2.4.1, we have extracted the conductance of the samples and analyzed it in comparison with the published data obtained by time-domain spectroscopy (TDTS) [21]. Our data allowed us to extract the optical conductivity and to compare it with that of the crystalline samples analyzed above. This comparison provides a good way to observe the degree of compensation, that is the insulating character of the bulk, in order to individuate the best sample to study the 2DEG and the topological surface carriers by IR spectroscopy.

3.2.1 Transmittance and Conductance

In Fig. 3.10 we report the transmittance data of both thin films in the sub-THz and THz regions from 5 K to 300 K. 60 and 120 QL films are reported in panel (a) and (b), respectively. The low-frequency data have also been measured with coherent synchrotron radiation in low-α mode (see Sect. 2.1.1), in order to better identify the free-carrier contribution. The α and β phonons are clearly visible at about 65 and 134 cm^{-1}, respectively, with the latter being better defined in the thicker sample. The most intense phonon (α-mode) surprisingly softens for decreasing T, while—as expected—the well defined Drude term narrows when T decreases.

[1] The upper limit for the measured frequency region is due to the transparency window of the substrate.

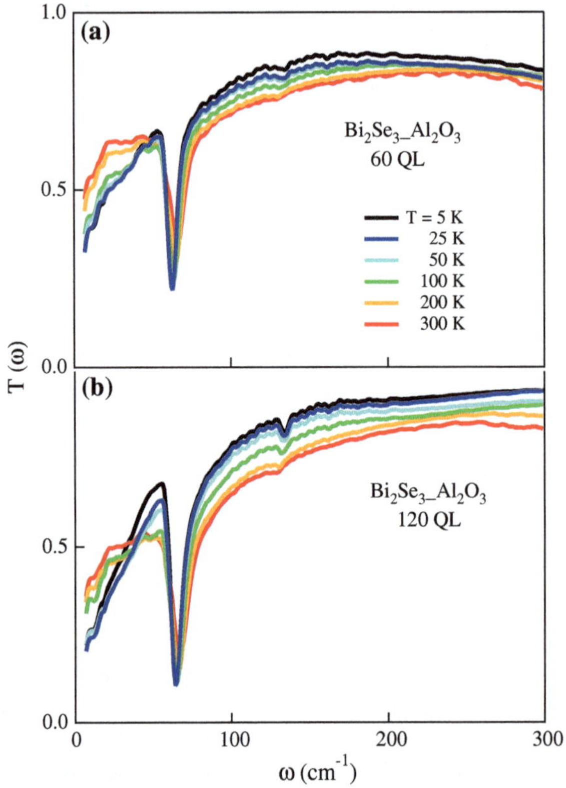

Fig. 3.10 Transmittance of two TI thin films of $Bi_2Se_3\_Al_2O_3$ at different temperatures in the sub-THz and THz regions: 60 QL thick (**a**) and 120 QL thick (**b**) (1 QL ~ 1 nm)

The conductance extracted from the transmittance by the procedure illustrated in Sect. 2.4.1 are shown in Fig. 3.11 between 0 and 150 cm^{-1} at the same temperatures. At first sight, one can easily see that both phonons are stronger in the thicker sample, providing evidence for their bulk character. They also broaden and harden with increasing temperature. At lower frequency, the free-carrier contribution (Drude term) does not change with thickness. Indeed, as one can see in the inset of Fig. 3.11, for a fixed temperature, the Drude term has the same width for both the samples.

This behavior suggests that, if a bulk feature, like a phonon mode, renormalizes with the thickness, the Drude term does not. This shows that the latter absorption originates from the surface of the sample, which doesn't change with the thickness, probably from the surface carriers in their topological surface state (see also Ref. [21]).

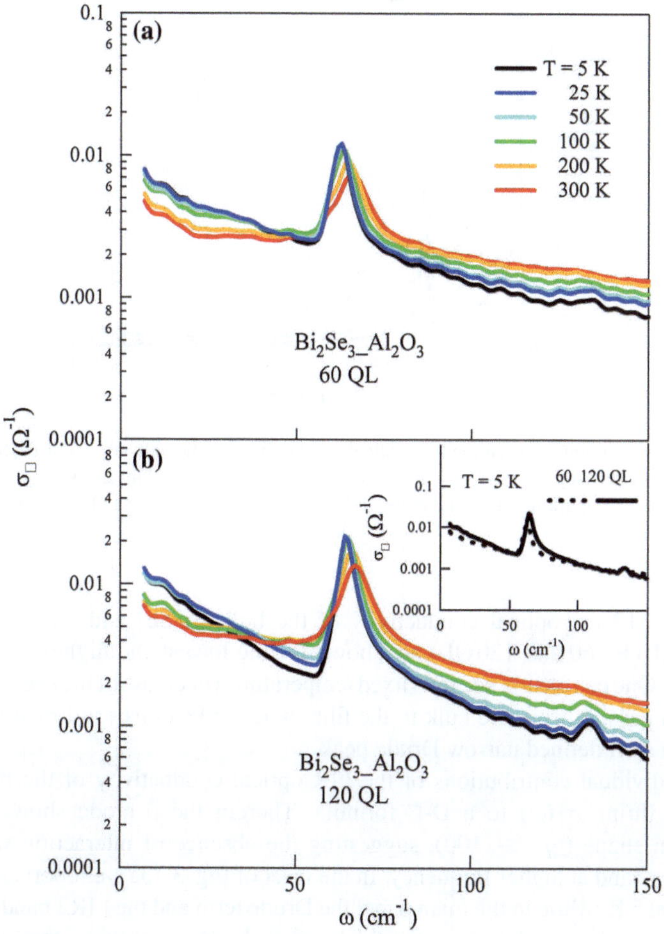

Fig. 3.11 Real part of the conductance of the two TI thin films of $Bi_2Se_3\_Al_2O_3$ at different temperatures in the sub-THz and THz regions: 60 QL thick (**a**) and 120 QL thick (**b**). In the inset a comparison between the real part of the conductance of the 60 QL thick film and the 120 QL thick one at 5 K is reported

3.2.2 Optical Conductivity

In order to better characterize the TI thin films in comparison with the TI crystals analyzed above, we report a comparison of their respective optical conductivities.

We have calculated the optical conductivity of the thin films simply considering that

$$\sigma_1(\omega) = \sigma_{\Box}(\omega)/t$$

where t is the thickness of the sample and $\sigma_{\Box}(\omega)$ is the conductance (see Sect. 2.4.1).

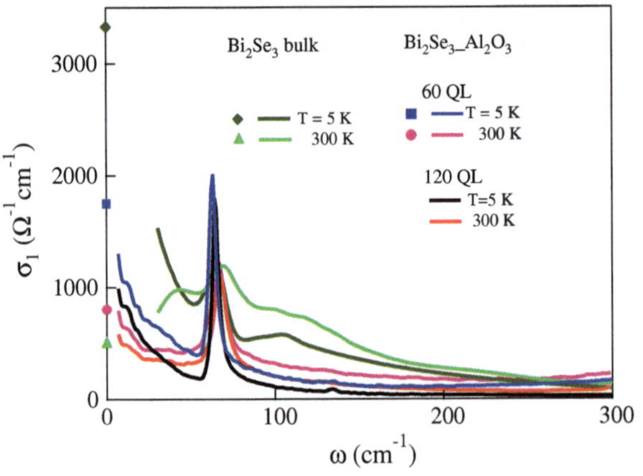

Fig. 3.12 Comparison between the FIR optical conductivity of the bulk crystal of Bi_2Se_3 and that of the thin films of $Bi_2Se_3\_Al_2O_3$ at the lowest and highest temperature measured. The symbols on the vertical axis mark the dc conductivity of samples from the same batch of the crystal and of similar films

In Fig. 3.12 the optical conductivity of the bulk Bi_2Se_3 and that of the films $Bi_2Se_3\_Al_2O_3$ (60 and 120 QL) are shown for the lowest and highest temperature measured. One can clearly see for a fixed temperature a decrease of the spectral weight SW when passing from the bulk to the film, where SW moves to low frequencies, showing a well defined narrow Drude peak.

The individual contributions of the FIR optical conductivity of the films were found by fitting $\sigma_1(\omega)$ to a D-L formula. Therein the α-mode shows a nearly Lorentzian shape ($|q| \gg 100$), suggesting the absence of interaction with some continuous band at higher frequency. In the inset of Fig. 3.13a we report an example of that fit at 5 K, while in the main panel the Drude term and the FIR2 band, centered at about 90 cm^{-1}, are shown for the thicker sample. In panel (b) of the same figure one can observe the comparison between the two FIR contributions in the bulk and in the film of the same material. The FIR2 band, associated to an impurity band (as described in details in Sect. 3.1.3) is strongly reduced in favor of a narrow Drude peak: this implies a predominance of the intrinsic contribution of surface carriers in the film, with respect to the extrinsic one due to the not perfect stoichiometry of the sample.

3.3 Spectral Weight in Thin Films and Crystals

A quantitative comparison between the charge density expected for the topological surface states and that provided by extrinsic charge carriers can be obtained by calculating the optical spectral weight $SW(\Omega)$, that is

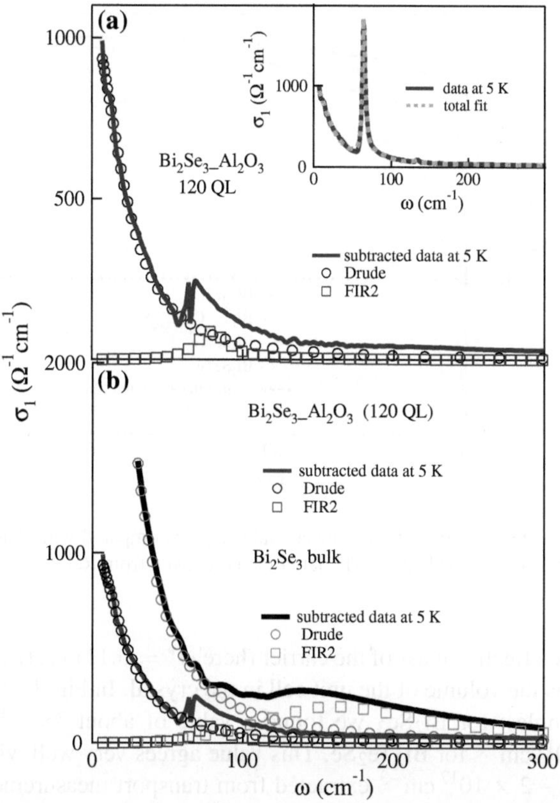

Fig. 3.13 FIR optical conductivity at 5 K of the thicker (120 QL) film of Bi_2Se_3_Al_2O_3 after subtraction of the phonon contributions via a D-L fit. The Drude term and the FIR2 contributions are indicated by open circles and squares, respectively. In the inset the fit (*dotted line*) and data at 5 K are shown (**a**). A comparison is proposed for the Drude term and FIR2, between the FIR-subtracted optical conductivities at 5 K of the Bi_2Se_3 crystal and that of the film (**b**)

$$SW(\Omega) = \int_0^{\Omega} \sigma_1^{sub}(\omega, T) d\omega \tag{3.1}$$

where $\Omega = 500\,\text{cm}^{-1}$ is a cutoff frequency, which well separates (see Fig. 3.4) the low-frequency excitations from the interband transitions. The integrated conductivity is the one subtracted of both the phonon and the interband contributions by a D-L fit. This ensures to analyze only the intraband extrinsic contributions. The spectral weight decreases by one order of magnitude from Bi_2Se_3 to Bi_2Te_2Se, showing the drastic effect of chemical compensation obtained through Te substitution [11]. From the SW one can estimate a 3D charge density $n_V(\Omega)$, according to the relation

$$\frac{m}{m^*} n_V(\Omega) = \frac{2mV}{\pi e^2} \int_0^{\Omega} \sigma_1^{sub}(\omega, T) d\omega \tag{3.2}$$

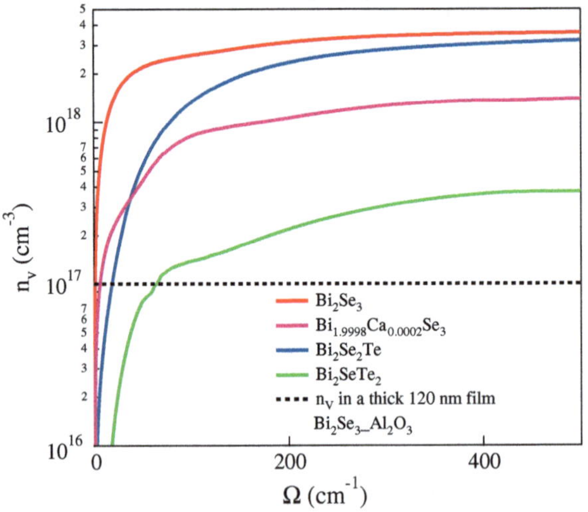

Fig. 3.14 Bulk carrier densities of the four crystal samples compared with that one of a Bi_2Se_3 thin film (120 nm) on Al_2O_3 substrate. The latter data are taken from Ref. [22]

where m^* is the effective mass of the carrier (here $m^* = 0.11\,\mathrm{m}$ [2]), e is the electron charge and V is the volume of the unit cell in the crystal. In Fig. 3.14 n_V is reported for all the samples: for Bi_2Se_3 we found a value of about $3\times10^{18}\,\mathrm{cm}^{-3}$, which becomes $\sim 10^{17}\,\mathrm{cm}^{-3}$ for Bi_2Te_2Se. This value agrees very well with the range of $5\times10^{16}\,\mathrm{cm}^{-3} - 2\times10^{17}\,\mathrm{cm}^{-3}$ extracted from transport measurements in crystals of the same batch [10] (see also Sect. 1.3.2).

In the same figure, we have also reported (dashed line) the bulk contribution to the carrier density of a thin film of Bi_2Se_3, belonging to the same batch of that one of our films. Such carrier density is estimated by the growers from the sheet carrier density, n_s, assuming that $n_V \sim n_s/t$, where t is the thickness of the sample [22] (see also Sect. 1.3.2). The estimated value is $n_V \sim 10^{17}\,\mathrm{cm}^{-3}$. The actual 3D charge density of the most compensated sample, Bi_2Te_2Se, calculated from its spectral weight, is still higher, by a factor of 3, than that associated with a thin film. In particular, we have estimated that the SW excess comes from the impurity bands located at finite (FIR) frequency. This result indicates that the low-energy electrodynamics in single crystals of TI, even at the highest degree of compensation presently achieved and the lowest temperatures where infrared spectra are taken, is still influenced by 3D charge excitations.Therefore, further improvements in the compensation are needed before bulk techniques, such as infrared spectroscopy, may observe, in single crystals, the optical properties of purely topological metallic states.

This is not true for TI thin films, where the absolute majority of SW corresponds to the Drude term, which describes the 2DEG response of surface states. The value of surface carrier density of those films, allowed us to investigate their 2DEG by the means of optical spectroscopy.

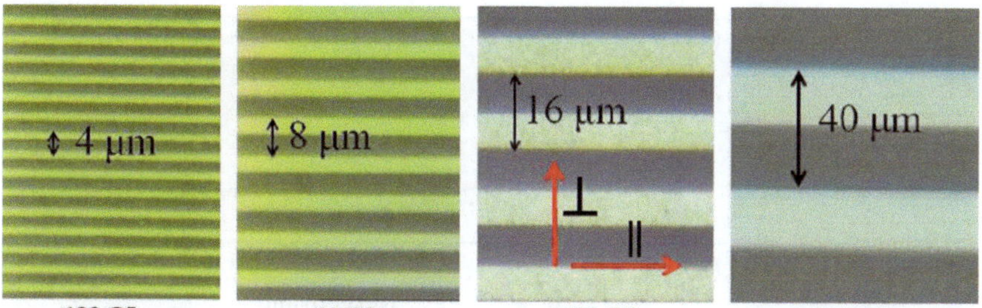

Fig. 3.15 Images at the electron microscope of the four films with gratings of different width; the *red arrows* show the direction of light polarization: perpendicular and parallel to the wires, respectively. The thickness of the films is reported under the images

3.4 Surface Plasmon Polaritons in Patterned Topological Insulator Thin Films

In this section we analyze the optical measurements of the four TI thin films, characterized in the previous section, patterned with a grating as described in Sect. 2.5.3.

In particular the samples are:

- Bi_2Se_3_Al_2O_3 120 QL thick, 40 μm grating period
- Bi_2Se_3_Al_2O_3 120 QL thick, 16 μm grating period
- Bi_2Se_3_Al_2O_3 60 QL thick, 8 μm grating period
- Bi_2Se_3_Al_2O_3 120 QL thick, 4 μm grating period

As shown in Fig. 3.15 the filling factor is the same for all samples (0.5). We have measured their transmittance in two light polarizations, in order to investigate SSPs excitations (see Sect. 2.4.2).Their observation may give an unambiguous signature of collective modes of surface charge carriers.

In Fig. 3.15 red arrows show the direction of polarization parallel to the wires of grating and the one perpendicular. As expected from the theory of SPPs (Sect. 2.4.2), they are observed only when the radiation field is polarized perpendicular to the wires.

3.4.1 Extinction Coefficient

We have measured the transmittance of the four samples in the FIR at different temperatures, from 6 to 300 K, and we have extracted the extinction coefficient $\epsilon(\omega)$ (Eq. 2.67). In Fig. 3.16 $\epsilon(\omega)$ is reported for both polarizations at different temperatures.

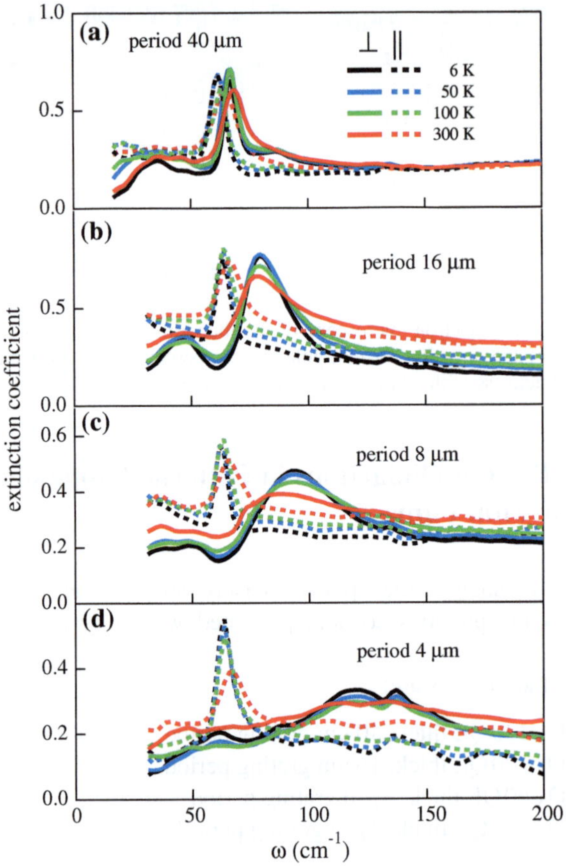

Fig. 3.16 Extinction coefficient of the four samples with different grating periods at various temperatures. *Dotted lines* refer to parallel, *solid lines* to perpendicular light polarization

One can clearly recognize in parallelly polarized data (dotted lines) the α phonon at about 61 cm^{-1} superimposed to a Drude term. Solid lines, referred to perpendicular polarized data, show different features at higher frequency with respect to the phonon one. These features, at first sight can be associated with surface plasmons. In order to better compare the spectra, we have normalized them to their respective peak values. The results are reported in Fig. 3.17 at three main temperatures. The dotted lines show even more clearly the phonons (α- and β- modes) superimposed to a decreasing versus frequency background (Drude term). Figure 3.18 shows the comparison at low temperature between the parallelly polarized data and those for the not patterned films, described in the previous section. The width of the main phonon is nearly the same and the narrowing of the Drude term is comparable. This ensures that the etching didn't affect the crystal structure nor the conductivity of the carriers.

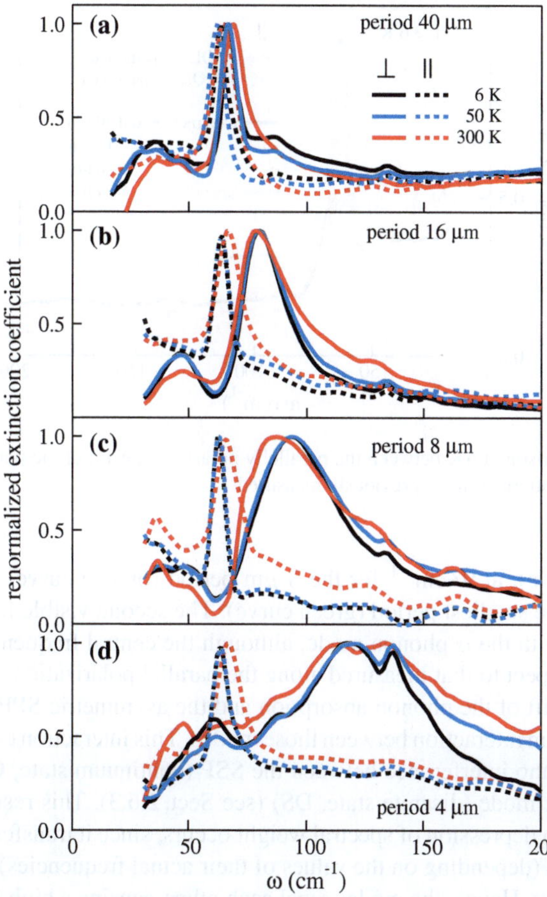

Fig. 3.17 Extinction coefficient of the four samples normalized to their respective peak values for convenience of comparison at 6, 50 e 300 K

The solid lines in Fig. 3.17, on the other hand, exhibit two main features. The first one, with a large, nearly-Lorentzian shape, moves from about 40 cm^{-1} to 120 cm^{-1}. Both its line shape and dependence on grating period indicate that it is indeed a surface plasmon, as its central frequency increases as the grating width decreases. The second feature strongly depends on grating width. It is quite narrow, intense and close to the α phonon in panel (a), not very intense, quite broadened and lower in frequency than the α phonon one in panels (b), (c) and (d).

In Fig. 3.19 the normalized extinction coefficient at 6 K is reported for parallel polarization in panel (a) and for the perpendicular one in panel (b). In the latter panel, the comparison between the four SPPs can be better appreciated. Their central frequency increase as the grating period decreases. In particular, the SP appears at about 36 cm^{-1} for the largest period (black curve), at 79 cm^{-1} for the 16 μm large

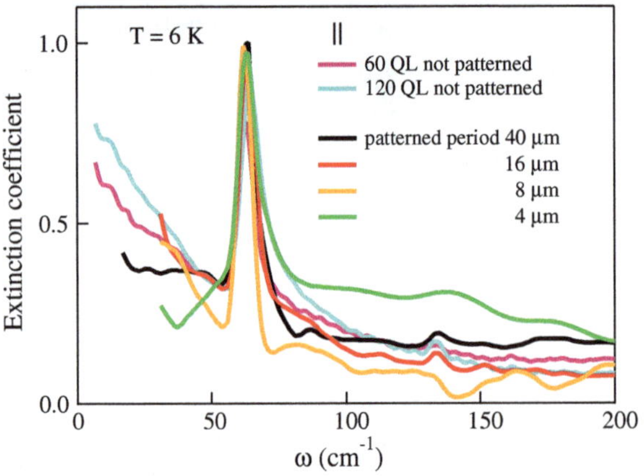

Fig. 3.18 Comparison at 6 K between the parallelly polarized spectra of the patterned films with those one of not patterned films, previously measured

period (red curve), at 93 cm^{-1} for the 8 μm period (orange curve) and, finally, at 119 cm^{-1} for the smallest period (green curve). The second visible feature seems to be associated with the α phonon mode, although the central frequency seems to be shifted with respect to that measured along the parallel polarization.

The huge shift of the phonon absorption and the asymmetric SPP shape suggest the presence of an interaction between those modes. This interaction can be described in terms of a Fano interference between the SSP (continuum state, CS) and the IR active phononic mode (discrete state, DS) (see Sect. 2.6.3). This results in a mixed state, in which a depression of spectral weight occurs, since it transfers from a mode to the other one (depending on the values of their actual frequencies), providing the shift of the peaks. Hence, the modes repel each other, causing a high frequency shift for the mode at high frequency and a softening for that at low frequency.

In other words, if ω_{bph} is the "bare" frequency of the phonon in the absence of the interaction (known from the spectra in panel (a)), it corresponds to the dip (or inflection point) of the mixed state in panel (b), where the spectra are the result of the excitation of both the DS and CS.

This scenario is a clear signature of a Fano resonance, that will be analyzed in further detail in the next section.

3.4.2 Fano Resonances in Plasmonic Absorption in TIs

In order to identify the unknown "bare" frequency of the SSPs and hence to study its behavior, we fit the normalized extinction coefficient (perpendicular polarization) to a function, described by Eqs. 2.78–2.81 in Sect. 2.6.3.

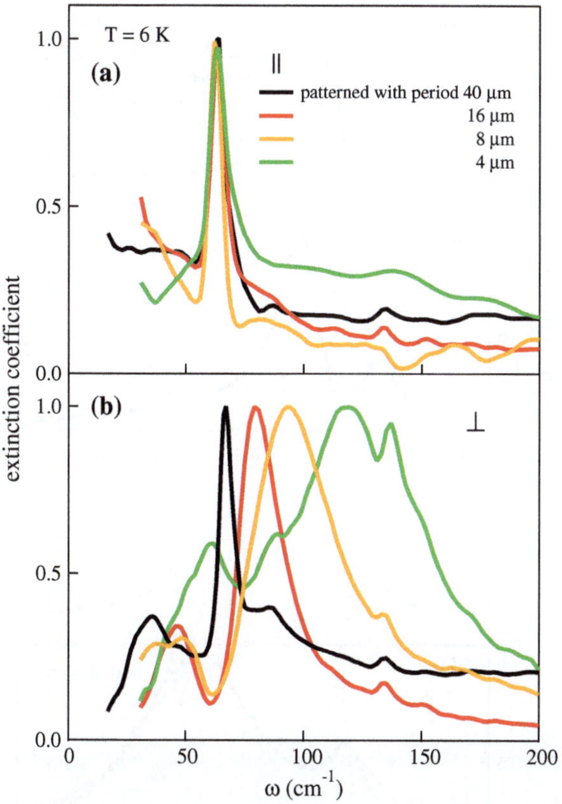

Fig. 3.19 Parallelly polarized (**a**) and perpendicularly polarized spectra (**b**) at 6 K for the four samples

In Fig. 3.20 we report the data at 6 K and the related fits (black dotted lines). In panel (a) the spectrum has a very similar aspect to that calculated and illustrated in Fig. 2.26b, since here we are in the regime $w \gg g$, because the plasmonic mode is less intense then the phononic one. In particular, the resonance occurs between a SSP and the α phonon, when the frequency of the former one is lower than that of the latter one. From the fits we have extracted the bare frequencies of the modes, whose peaks are shown in Fig. 3.20 (colored dotted lines). Both phonon peaks recover those observed in the parallelly polarized spectra. It is very clear how the interaction leads the two peaks (SPP and α-mode) to move away from each other. In particular the bare plasmon frequency is shifted to the higher frequencies with respect to the peak of the mixed state.

In panels (b) and (c) the situation is different. The spectra recover the calculated ones of the Fig. 2.26e, in which the coupling factor of the phonon is less then the plasmon one (regime $w \ll g$). Here, the frequency of the bare phonon is lower then

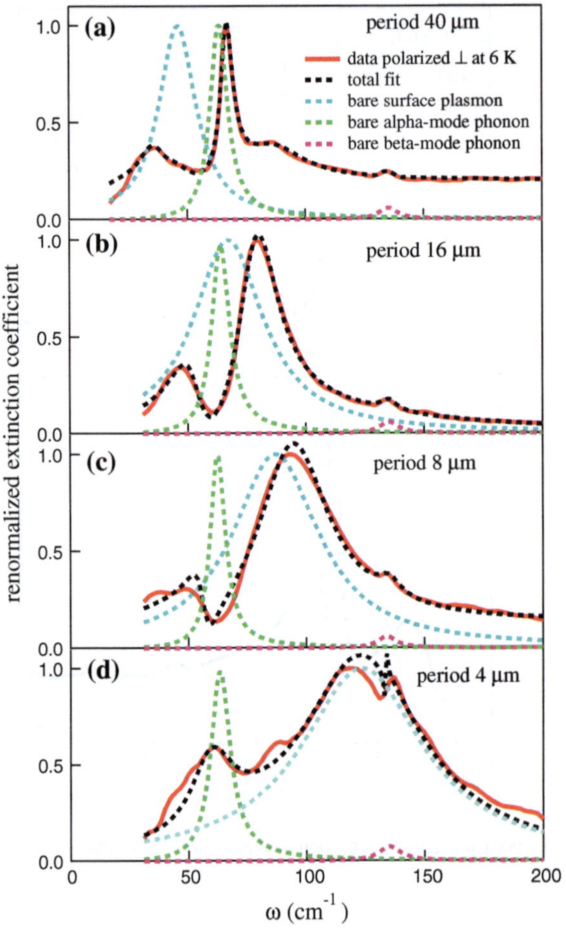

Fig. 3.20 Perpendicularly polarized spectra of the four samples at 6 K. The fits to data are reported (*black dotted lines*) together with the reconstructed, bare plasmons and phonons (*colored dotted lines*)

the SPP one, so that the bare plasmon peak is shifted to lower frequency with respect to the mixed state.

In panel (d) the situation is now that illustrated in Fig. 2.26a, since the interference occurs between the SPP and the β phonon. This regime is again that in which the coupling factor of the DS is less then the CS one ($w \ll g$), being the phonon intensity very small. Here the frequency of the bare plasmon is higher then the one of the mixed state. On the contrary the peak of the α-mode recovers quite well the bare one.

The main fit parameters for all the samples are reported in Table 3.2. There are the frequencies of the bare plasmon and the bare phonon, which interact each other, respectively. One can thus quantitatively appreciate the frequency shift from the

Table 3.2 Fitting parameters of the data shown in Fig. 3.20

	v	bare surface plasmon ω_{bp} (cm^{-1})	bare interacting phonon ω_{bph} (cm^{-1})
120 QL 40 μm period	2.761	46 ±1	63 ±1 (α)
120 QL 16 μm period	2.069	67 ±1	64 ±1 (α)
60 QL 8 μm period	2.034	87 ±1	63 ±1 (α)
120 QL 4 μm period	0.189	123 ±1	135 ±1 (β)

v is the coupling between the SSP and the phonons. The frequencies of bare SSP and bare interacting phonon are reported

mixed state. Moreover, one can see how the coupling between the CS and DS does not change too much, as the period varies, for the interaction between the SPP and the most intense phonon (α-mode), while it decreases by a factor of 10 when, in the fourth line, the interaction is with the weakest phonon (β-mode).

3.4.3 Plasmonic Dispersion

Once we have found the bare frequency of the SPPs excited in TI thin films, we have analyzed its scaling both with the width of grating w and the SPP wavevector k, that is related to the period of grating a by the relation $k = 2\pi/a$.

The SPP frequency for an electronic 2DEG has to scale with $\sqrt{k}$, that is what we show in Fig. 3.21. The four bare plasmon frequencies (see Table 3.2) are reported (red symbols). They are fitted to a power law, $\omega = \mathrm{A}k^{1/2}$ (dotted line), where A is a parameter. The fit is in very good agreement with data, strongly suggesting a 2D nature of charge carriers.

Such a behavior is moreover independent from the temperature. In Fig. 3.22 we report the bare plasmon frequencies at 6 K together with those extracted by the same fits, described above, to the spectra at 50 K (see Fig. 3.17), and not shown. It is clear that the two groups of data follow the same power law.

Moreover, ω_{bp} follows a linear dependence on $\mathrm{w}^{-1/2}$ (inset of Fig. 3.21). This is another strong indication of a 2DEG behavior. However, those plasmonic excitations cannot be associated only with 2D Dirac charge carriers. Indeed, it is well known that in small gap insulators bend bending phenomenon (due to electrostatic effect at the surface) may change the dispersion of bulk valence band [22], inducing metallic (extrinsic) states at the interface between those insulators and the vacuum. This generates a metallic conduction at the surface, i.e. a parallel channel with respect to the intrinsic Dirac channel. Both carrier channels show a 2D $\sqrt{k}$ dispersion in agreement with experimental data in Fig. 3.21.

A specific signature of Dirac contribution to plasmonic excitations comes from a $n^{1/4}$ dependence (where n is the surface charge density), at variance with a more conventional $n^{1/2}$ in massive electrons [23].

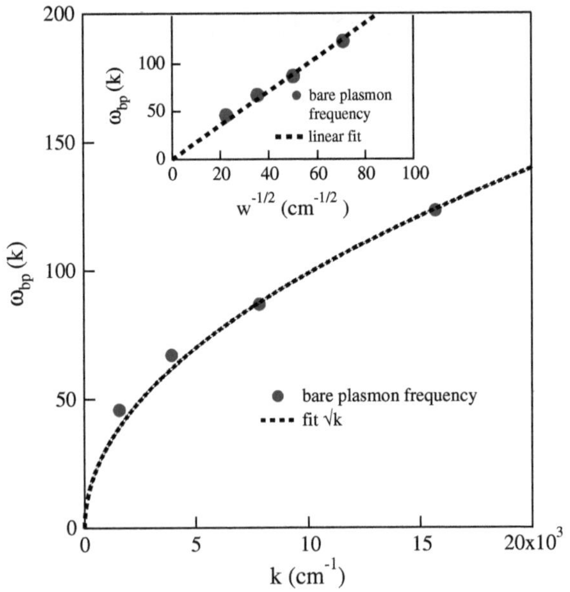

Fig. 3.21 Dispersion $\omega_{bp}(k)$ at 6 K. The *dotted line* is a $\sqrt{k}$ fit of data. In the inset the scaling of $\omega_{bp}(k)$ with $w^{-1/2}$ is also shown, with its linear fit (*dotted line*). The error bars are within the symbol size

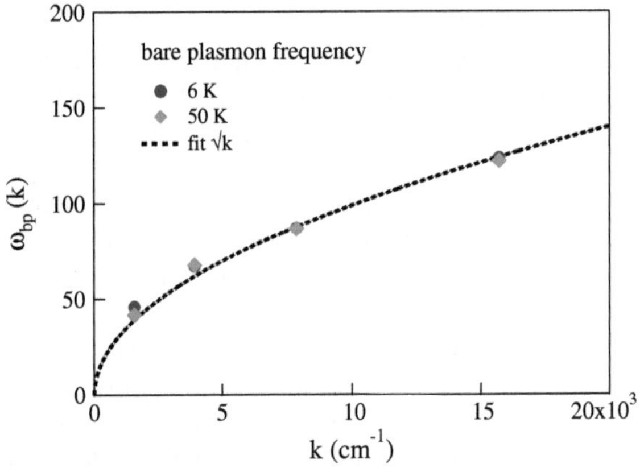

Fig. 3.22 Dispersion $\omega_{bp}(k)$ and its related fit shown at two different temperatures: 6 and 50 K. The error bars are within the symbol size

A perspective of this work may be therefore to implement a patterned device with TIs samples, where one can tune the carrier density, in order to investigate the excitation of SSP's as a function of n.

3.5 Conclusions

This thesis work illustrates an experimental investigation, performed by optical spectroscopy, of Topological Insulator materials in the sub-THz and IR ranges.

We have first performed reflectivity measurements on four TI crystals (Bi_2Se_3, $Bi_{1.9998}Ca_{0.0002}Se_3$, Bi_2Se_2Te and Bi_2Te_2Se) at different temperatures. By means of FTIR spectroscopy we have extracted and analyzed their optical conductivity, in order to understand how chemical doping and chemical compensation affect the extrinsic contribution of charge carriers.

We have observed in the optical conductivity the presence of both a well defined Drude term and a FIR absorption in the non compensated sample (Bi_2Se_3). In the $Bi_{1.9998}Ca_{0.0002}Se_3$ and Bi_2Se_2Te the increased chemical compensation and the consequent reduction in the carrier density results in an overall decrease of the FIR spectral weight. Moreover the FIR absorption splits into two bands, labelled FIR1 and FIR2. This double spectral structure is finally also observed in the most compensated sample (Bi_2Te_2Se), where the Drude term is strongly suppressed. In analogy with what has been observed in other conventional doped semiconductors, such as Si:P, the FIR2 band at 200 cm^{-1} can be assigned to the transitions from the impurity bound states to the electronic continuum. This band is also in good agreement with the impurity ionization energy estimated from the temperature dependence of resistivity and Hall data, namely 20–40 meV. Instead, the low-frequency band FIR1 at 50 cm^{-1} can be associated with hydrogen-like $1s \rightarrow np$ transitions, broadened by the inhomogeneous environment of the impurities and/or by their interaction.

Furthermore, the phonon structure of the samples has been analyzed. Two IR-active phonons have been observed, one (α-mode) at about 61 cm^{-1} and the other (β-mode) at about 133 cm^{-1}. Since the former exhibits a clear asymmetric line shape, we have fitted the optical conductivity to a Drude-Lorentz model with a Fano line shape for the α-mode. This Fano character suggests an interaction between the phonon and the electronic continuum corresponding to the FIR1 band.

The spectral weight of single crystals and the corresponding 3D charge density, even in the most compensated sample, is still higher than the spectral weigth related to surface carriers. This demonstrates that the extrinsic bulk conductivity masks the topological surface properties even in the most compensated single crystal. Hence, we have focused our interest on the IR spectroscopy of TI thin films, where the ratio of surface contribution to that of the bulk is more favorable. We have measured the FIR transmittance on Bi_2Se_3 thin films on sapphire substrate of two different thickness (60 and 120 nm). Once extracted their conductance, we have revealed the nearly independence to the thickness of free carrier contribution. This have suggested that the conducting contribution comes from the surface, i.e. from the electronic 2DEG, providing to us, for the first time in our knowledge, the possibility to investigate by optical spectroscopy the topological plasmonic excitations.

This has been done fabricating patterned TI thin films with four different period of grating. Such devices allowed us to reveal the excitation of collective modes of the surface charge density (called surface plasmons polariton, SPPs), associated with the 2DEG.

The surface plasmon polariton frequency here observed follow a linear dependence with the square root of the grating wavevector. This is in perfect agreement with a 2DEG behavior.

As it is well known that in small gap insulators bend bending phenomenon may change the dispersion of bulk valence band, inducing metallic (extrinsic) states at the interface between those insulators and the vacuum, not only Dirac charge carriers contribute to the 2DEG of a TI. Therefore, a perspective of this work is the quest of a clear signature of Dirac carriers contribution to plasmonic excitations. Indeed, one can calculate the theoretical dispersions due to both Dirac (non massive) carriers and massive carriers (due to the band bending effect) from experimental values of their charge densities, Fermi velocity and mass. This is what we have done beyond this thesis and reported in Ref [24].

Another close perspective about the experimental evidence of Dirac plasmons in TIs is measuring the plasmonic dispersion as a function of the carrier density n_D. Indeed, a $n_D^{1/4}$ dependence, at variance with a more conventional $n^{1/2}$ in massive electrons, could definitely prove that the plasmonic excitation is due to Dirac carriers. This can be done implementing a patterned device with TIs samples, where one can tune the carrier density, for instance by an electrostatic gating.

Finally, a further perspective of this thesis could be exploiting the dynamical behavior of the plasmons in TIs using the pump and probe technique in the THz range.

References

1. D.-X. Qu, Y.S. Hor, J. Xiong, R.J. Cava, N.P. Ong, Science **329**, 821 (2010)
2. A.D. LaForge, A. Frenzel, B.C. Pursley, T. Lin, X. Liu, J. Shi, D.N. Basov, Phys. Rev. B **81**, 125120 (2010)
3. N.P. Butch, K. Kirshenbaum, P. Syers, A.B. Sushkov, G.S. Jenkins, H.D. Drew, J. Paglione, Phys. Rev. B **81**, 241301(R) (2010)
4. Y.S. Hor, A. Richardella, P. Roushan, Y. Xia, J.G. Checkelsky, A. Yazdani, M.Z. Hasan, N.P. Ong, R.J. Cava, Phys. Rev. B **79**, 195208 (2009)
5. E.M. Black, E.M. Conwell, L. Seigle, C.W. Spence Phys, Chem. Sol. **2**, 240 (1957)
6. W. Richter, H. Köler, C.R. Becker, Phys. Stat. Sol. (b) **84**, 619 (1977)
7. H. Zhang, C.X. Liu, X.L. Qi, X. Dai, Z. Fang, S.C. Zhang, Nat. Phys. **5**, 438 (2009)
8. Z. Ren, A.A. Taskin, S. Sasaki, K. Segawa, Y. Ando, Phys. Rev. B **82**, 241306(R) (2010)
9. J. Xiong, Y. Luo, Y. Khoo, S. Jia, R.J. Cava, N.P. Ong, Physica E: Low-dimensional Systems and Nanostructures **44**(5) 920 (2012)
10. Shuang Jia Huiwen Ji, E. Climent-Pascual, M.K. Fuccillo, M.E. Charles, J. Xiong, N.P. Ong, R.J. Cava, Phys. Rev. B **84**, 235206 (2011)
11. P. Di Pietro, F.M. Vitucci, D. Nicoletti, L. Baldassarre, P. Calvani, R. Cava, Y.S. Hor, U. Schade, S. Lupi, Phys. Rev. B **86**, 045439 (2012)
12. D.L. Greenaway, G. Harbeke, J. Phys. Chem. Solids **26**, 1585 (1965)
13. R. Vilaplana, D. Santamaría-Pérez, O. Gomis, F. J. Manjón, J. González, A. Segura, A. Muñoz, P. Rodríguez-Hernández, E. Pérez-Gonzéalez, V. Marín-Borrás, V. Muñoz-Sanjose, C. Drasar, V. Kucek, Phys. Rev. B **84**, 184110 (2011)

14. H. Köhler, C.R. Becker, Physica status solidi (b) **61**(2), 533 (1974)
15. A. Gaymann, H.P. Geserich, H.V. Lohneysen, Phys. Rev. B **52** 16486 (1995)
16. G.A. Thomas, M. Capizzi, F. DeRosa, R.N. Bhatt, T.M. Rice Phys, Rev. B **23**, 5472 (1981)
17. N.F. Mott, *Metal-insulator transitions* (Taylor and Francis, London, 1990)
18. Y.S. Kim, M. Brahlek, N. Bansal, E. Edrey, G.A. Kapilevich, K. Iida, M. Tanimura, Y. Horibe, S.-W. Cheong, S. Oh, Phys. Rev. B **84**, 073109 (2011)
19. A. Damascelli, K. Schulte, D. van der Marel, A.A. Menovsky, Phys. Rev. B **55**, R4863 (1997)
20. S. Lupi, M. Capizzi, P. Calvani, B. Ruzicka, P. Maselli, P. Dore, A. Paolone, Phys. Rev. B **57**, 1248 (1998)
21. R. Valdés Aguilar, A.V. Stier, W. Liu, L.S. Bilbro, D.K. George, N. Bansal, L. Wu, J. Cerne, A.G. Markelz, S. Oh, N.P. Armitage, Phys. Rev. Lett. **108**, 087403 (2012)
22. N. Bansal, Y.S. Kim, M. Brahlek, E. Edrey, S. Oh, Phys. Rev. Lett. **109**, 116804 (2012)
23. L. Ju, B. Geng, J. Horng, C. Girit, M. Martin, Z. Hao, H.A. Bechtel, X. Liang, A. Zettl, Y.R. Shen, F. Wang, Nat. Nanotechnol. **6**, 630 (2011)
24. P. Di Pietro, M. Ortolani, O. Limaj, A. Di Gaspare, V. Giliberti, F. Giorgianni, M. Brahlek, N. Bansal, N. Koirala, S. Oh, P. Calvani, S. Lupi, Nat. Nanotechnol. **8**, 556–560 (2013)
25. E. Janzén, R. Stedman, G. Grossmann, H.G. Grimmeiss, Phys. Rev. B **29**, 1907 (1984)
26. R. Valdés Aguilar, A.V. Stier, W. Liu, L.S. Bilbro, D.K. George, N. Bansal, L. Wu, J. Cerne, A.G. Markelz, S. Oh, N.P. Armitage, Supplemental Materials for: Phys. Rev. Lett. **108**, 087403 (2012)

MIX
Papier aus verantwortungsvollen Quellen
Paper from responsible sources
FSC® C105338
FSC
www.fsc.org

If you have any concerns about our products,
you can contact us on
ProductSafety@springernature.com

In case Publisher is established outside the EU,
the EU authorized representative is:
Springer Nature Customer Service Center GmbH
Europaplatz 3, 69115 Heidelberg, Germany

Printed by Libri Plureos GmbH
in Hamburg, Germany